AF388434

Mohammad Assadsolimani

Higher order QCD corrections to single top quark production

disserta
Verlag

Assadsolimani, Mohammad: Higher order QCD corrections to single top quark production, Hamburg, disserta Verlag, 2014

Buch-ISBN: 978-3-95425-674-7
PDF-eBook-ISBN: 978-3-95425-675-4
Druck/Herstellung: disserta Verlag, Hamburg, 2014

Bibliografische Information der Deutschen Nationalbibliothek:
Die Deutsche Nationalbibliothek verzeichnet diese Publikation in der Deutschen
Nationalbibliografie; detaillierte bibliografische Daten sind im Internet über
http://dnb.d-nb.de abrufbar.

Zugl.: Berlin, Humboldt-Univ., Diss., 2014

Higher order QCD corrections to single top quark production

DISSERTATION

zur Erlangung des akademischen Grades

doctor rerum naturalium
(Dr. rer. nat.)
im Fach Physik

eingereicht an der
Mathematisch-Naturwissenschaftlichen Fakultät
der Humboldt-Universität zu Berlin

von
Dipl.-Physiker, Mohammad Assadsolimani
05.09.1981, Esfahan (Iran)

Präsident der Humboldt-Universität zu Berlin:
Prof. Dr. Jan-Hendrik Olbertz

Dekan der Mathematisch-Naturwissenschaftlichen Fakultät:
Prof. Dr. Elmar Kulke

Gutachter:

1. Prof. Peter Uwer

2. Prof. Dirk Kreimer

3. Prof. Janusz Gluza

Tag der mündlichen Prüfung: 18.07.2014

There is no quantum world. There is only an abstract physical description.
It is wrong to think that the task of physics is to find out how nature is.
Physics concerns what we can say about nature.

Niels Bohr

Abstract

It is known that the LHC has a large discovery potential because of the large center-of-mass energy ($\sqrt{s}$ =14 TeV) and the high design luminosity. In addition, the two experiments ATLAS [1] and CMS [2] perform precision measurements in numerous physics channels. The increasing experimental precision demands an even higher precision from the theoretical side. For a more precise prediction one has to consider the corrections obtained typically from Quantum Chromodynamics (QCD). The calculation of these corrections in the high energy regime is described within perturbation theory. In this work, multi-loop calculations in QCD and in particular two-loop corrections for single top quark production are considered. There are several phenomenological motivations to study single top quark production: Firstly, the process is sensitive to the electroweak Wtb–vertex and non–standard couplings can give a hint on physics beyond the Standard Model. Secondly, the t–channel cross section measurement provides information on the b–quark Parton Distribution Functions (PDF). Finally, single top quark production enables us to measure directly the Cabibbo-Kobayashi-Maskawa (CKM) matrix element V_{tb}. The next-to-next-to-leading-order (NNLO) calculation of the single top quark production has many building blocks. In this thesis two blocks will be presented: one-loop corrections squared and two-loop corrections interfered with Born. At first, the one-loop squared contribution at NNLO for single top quark production will be calculated. Before we begin with the calculation of the two-loop corrections to single top quark production, we calculate the QCD form factors of heavy quarks at NNLO and as a first independent check the axial vector coupling. A comparison with the literature [3] shows a complete agreement. This consistency check provides a proof for the validity of our setup. In the next step the two-loop corrections to single top quark production will be calculated. After reducing all occurring tensor integrals to scalar integrals, we apply the integration by parts method (IBP) [4] to find the master integrals. This step is a big challenge compared with all other similar calculations because of the number of variables in the problem (two Mandelstam variables s and t , the dimension d and the mass of the top quark m_t as well as the mass of the W boson m_w). Finally, the calculation of the three kinds of topologies, vertex corrections, double boxes and non-planer double boxes, in the two-loop contribution at NNLO calculation will be presented.

Zusammenfassung

Der LHC hat wegen der großen Schwerpunktsenergie ($\sqrt{s}$ =14 TeV) und einer hohen Luminosität ein großes Entdeckungspotential. Darüber hinaus sind die beiden Experimente ATLAS [1] und CMS [2] in der Lage für viele Prozesse Präzisionsmessungen durchzuführen. Das verlangt nach mehr Genauigkeit von theoretischer Seite. Um eine genauere Vorhersage machen zu können, muss man die Quantenkorrekturen betrachten. Für die meisten Prozesse haben die Quantenchromodynamik-Korrekturen (QCD-Korrekturen) einen dominanten Beitrag. Die Berechnung dieser Korrekturen im hochenergetischen Bereich wird durch die Störungstheorie beschrieben. In dieser Arbeit werden die Methoden für die Berechnung der Zweischleifen-Korrekturen in QCD beschrieben, jedoch das Hauptziel dieser Arbeit ist die Berechnung der Zweischleifen-Korrekturen für die Produktion der einzelnen Top-Quarks in nächst-zu-nächstführender Ordnung (NNLO) der QCD. Es gibt mehrere phänomenologische Gründe die Produktion einzelner Top-Quarks zu studieren: Erstens der Prozess ist empfindlich auf den elektroschwachen Wtb-Vertex, so dass Nicht-Standard-Kopplungen einen Hinweis auf die Physik jenseits des Standardmodells geben können. Zweitens bietet die Messung des t-Kanal-Wirkungsquerschnitts Informationen über die b-Quark-Partonverteilungsfunktion. Schließlich ermöglicht uns die Produktion einzelner Top-Quarks eine direkte Messung des Cabibbo-Kobayashi-Maskawa (CKM) Matrix-Elements V_{tb}. Die NNLO-Berechnung der Produktion der einzelnen Top-Quarks hat mehrere Bausteine. In dieser Arbeit werden zwei dieser Bausteine, Einschleifen-Korrekturen zum Quadrat und die Zweischleifen-Korrekturen, diskutiert. Zunächst werden Einschleifen-Korrekturen zum Quadrat in NNLO für die Produktion einzelner Top-Quarks berechnet. Bevor mit der Berechnung der Zweischleifen–Korrekturen begonnen wird, werden die QCD Formfaktoren schwerer Quarks in NNLO und als erste unabhängige Überprüfung, die axiale Vektor-Kopplung berechnet. Der Vergleich mit der Literatur [3] liefert eine vollständige Übereinstimmung. Dieser unabhängige Check liefert uns eine Sicherheit dafür, dass unser Setup richtig ist. In dem nächsten Schritt werden die Zweischleifen-Korrekturen für die Produktion einzelner Top-Quarks berechnet. Nachdem man alle auftretenden Tensor-Integrale zu skalaren Integralen reduziert hat, wird die sogenannte Integration-By-Parts-Methode [4] angewendet, um die Master-Integrale zu finden. Dieser Schritt ist ziemlich schwierig im Vergleich zu bislang durchgeführten Berechnungen aufgrund der Anzahl der Variablen, die in dem Problem auftauchen (zwei Mandelstam-Variablen s und t, die Dimension d und die Masse vom Top-Quark m_t sowie die Masse des W-Bosons m_w). Zum Schluss wird die Berechnung der drei Arten von Topologien: Vertex-Korrekturen, Doppelboxen und nicht-planare Doppelboxen, in dem Zweischleifen-Beitrag der NNLO-Rechnung präsentiert.

Contents

List of Figures

List of Tables

_1_

Introduction

The Standard Model (SM) of particle physics describes interactions of the fundamental constituents of all visible matter as well as the results of myriads of accelerator experiments. The discovery of a candidate for the Higgs boson at the LHC in 2012 [6, 7] has made an important step towards the experimental confirmation of the SM.

The Standard Model is a quantum field theory that collects our understanding of three of the four fundamental forces, i.e., the electromagnetic, weak and strong interactions as represented in tab. 1.1. The fourth force, gravity, is indeed classically understood, but is not described as a quantum field theory.

Forces:	Weak	Electromagnetic	Strong
Carried by:	W^+,W^-, Z	Photon	Gluon
Acts on:	Photon,Quarks, Leptons	Quarks, Leptons,W^+,W^-	Quarks, Gluons

Table 1.1.: The fundamental forces

The SM has two basic components: the spontaneously broken $SU(2) \times U(1)$ electroweak theory and the $SU(3)$ colour gauge theory, Quantum Chromodynamics (QCD), which describes the strong interaction. The strong interaction binds quarks and gluons into hadrons as well as protons and neutrons into the atomic nucleus.

To describe physical processes in the SM, we are interested in the calculation of a complex valued quantity, the so-called S-matrix. From the S-matrix one can calculate the scattering cross section for a certain scattering process, which can be compared with the experimental results. However, it is not possible to calculate the scattering cross section exactly. Therefore, one expands it in a perturbation series in a small parameter of the theory. The first term of this series is called the leading order and the other terms can be considered as corrections to the first term. The most relevant corrections typically come from QCD.

QCD shows a specific property in the high energy regime, namely that the interaction between quarks and gluons becomes progressively weaker. This property is called *asymptotic freedom* [8] and provides the conceptual framework for applying perturbative QCD to scattering cross sections at large-momentum-transfer. So we can apply this perturbation approach to a process if we are in the high energy regime because then the parameter in

1

which we want to expand the cross section is small. This parameter is the strong coupling constant α_s. In collision experiments when two hadrons collide at high energy, we are interested in the cross section which can be written as a factorized product of short and long distance processes [9]

$$\sigma(p_1, p_2) = \sum_{i,j} \int_0^1 dx_1 \int_0^1 dx_2 \; f_i(x_1, \mu_F^2) f_j(x_2, \mu_F^2) \hat{\sigma}_{i,j}(\mu_r^2, \mu_F^2), \tag{1.1}$$

where p_1 and p_2 are the momenta of the incoming hadrons as illustrated in fig. 1.1 and $\hat{\sigma}_{i,j}$ denotes the short distance cross section for the scattering of partons i and j, the so-called partonic cross section. The functions $f_i(x_1, \mu_F^2)$ and $f_j(x_2, \mu_F^2)$ are the QCD quark or gluon Parton Distribution Functions (PDFs), defined at a factorization scale μ_F which separates long and short distance physics [10]. μ_r is the renormalization scale. For the calculation

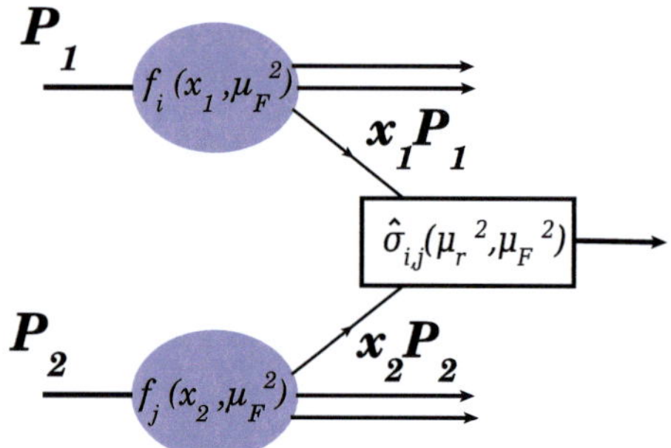

Figure 1.1.: The parton model description of a hard scattering process

of $\sigma(p_1, p_2)$ in eq.(1.1) we need two ingredients: the first one, the parton distribution functions $f_i(x_j, \mu_F^2)$ which are determined by experiments but their evolution with μ_F can be obtained by theory, and the second one, the partonic cross section $\hat{\sigma}$ expanded in the coupling constant α_s, which is calculated theoretically. The partonic cross section $\hat{\sigma}_{i,j}$ can be calculated as a perturbation series in the coupling constant α_s because it is small at high energy

$$\hat{\sigma} = \sigma^{\text{Born}} \left(1 + \frac{\alpha_s}{2\pi}\sigma^{(1)} + (\frac{\alpha_s}{2\pi})^2\sigma^{(2)} + (\frac{\alpha_s}{2\pi})^3\sigma^{(3)} + \cdots \right). \tag{1.2}$$

The lowest order in this series, σ^{Born}, is called leading order (LO) prediction. The next order, $\frac{\alpha_s}{2\pi}\sigma^{(1)}$, is called next-to-leading-order (NLO), the next one in the series, $(\frac{\alpha_s}{2\pi})^2\sigma^{(2)}$, is called next-to-next-to-leading-order (NNLO) and so on. Including higher corrections improves the predictions and decreases the scale dependence [11]. Going to higher order in the perturbation theory improves the matching between the parton level and the hadron level jet algorithm and reduces the error of α_s [12]. The higher the taken order is, the more complicated is the calculation of the series. Therefore, there are various techniques for the

calculation of NLO and higher order which differ principally. The techniques which can
be used at NNLO and higher order can be generally applied to NLO but not always vice
versa.

An NNLO calculation for a $2 \to 2$ scattering process has the following building blocks,
schematically

- two-loop corrections interfered with Born

- one-loop corrections squared

- one-loop corrections interfered
  with real corrections

- real corrections squared

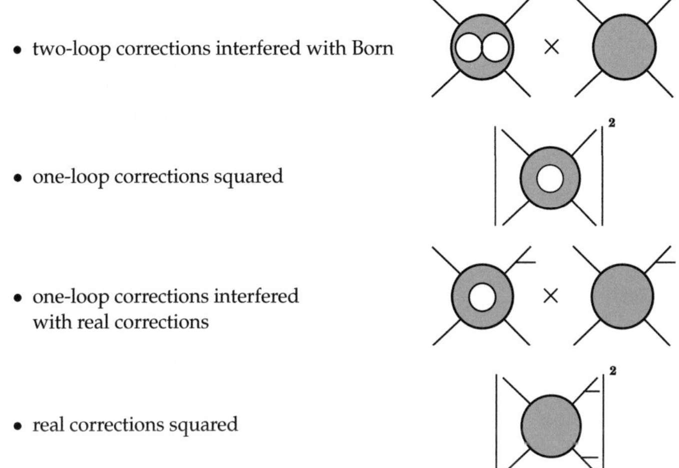

Figure 1.2.: Building blocks of an NNLO calculation for a $2 \to 2$ scattering process

We generally need an NNLO calculation of the series when the NLO corrections are
large, for instance in Higgs production, [13] or when truly high precision is needed, which
is often the case. Sometimes we have a rather good theoretical precision at NLO for the
cross section of a given process, in other words the deviation between the first and the
second term in eq.(1.2) is small, for instance *single top quark production* [14]. But we have to
consider that some colour structures do not contribute to the cross section at NLO because
of colour conservation, namely all one-loop diagrams involving the exchange of a single
gluon between the two quark lines. For example:

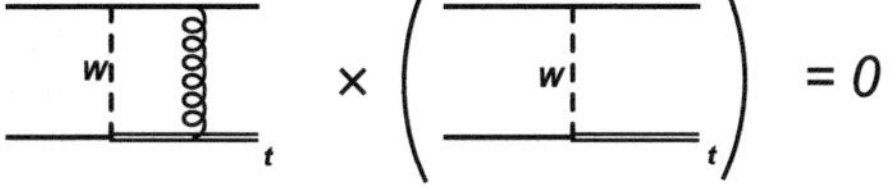

$$\quad w \quad t \quad \times \left(\quad w \quad t \quad \right)^* = 0$$

Figure 1.3.: Colour structure at NLO

However, these colour structures do contribute to the cross section at NNLO. Therefore, a reliable estimation of the precision can only be obtained by calculating the NNLO corrections.

The top quark in single top quark production is produced by the electroweak interaction. In the SM the three main production reactions are:

1. t-channel: $q\,b \to q'\,t$

2. s-channel: $q\,\bar{q}' \to \bar{b}\,t$

3. $t - W$ production mode: $b\,g \to W\,t$

There are several phenomenological reasons that motivate a high precision calculation of the single top quark production. First of all, it enables us to study the $V - A$ structure of the weak interaction directly [15, 16]. The next reason to study single top quark production is that the t-channel cross section measurement can be used to extract the b-quark PDF. Another important point is that the single top quark production plays an important role as a background for a number of Higgs search channels [17, 18] and new physics [19–21]. For instance, a new physics scenario involving flavor-changing neutral current (FCNC) would lead to single top production via $ug \to t$ or $cg \to t$ [22, 23]. One further reason is that because of the production mechanism, the produced single tops are highly polarized. Finally, single top quark production allows a direct measurement of the Cabibbo-Kobayashi-Maskawa (CKM) matrix element $|V_{tb}|$, which is known indirectly from unitarity

$$|V_{ub}|^2 + |V_{cb}|^2 + |V_{tb}|^2 = 1$$

to a very high precision, but not through a direct measurement. This indirect determination assumes three generations of quarks and the unitarity of the CKM matrix.

For the calculation of the loop contributions we have to deal with the so-called tensor integrals. A loop integral, in which the loop momenta occur also in the numerator is called a tensor integral and if the numerator of the integral is independent of the loop momenta, the integral is called a scalar integral. Every tensor integral should be reduced to a sum of many scalar integrals with different powers of the propagators. After that, one can evaluate each integral individually or alternatively express every one as a linear combination of some basic integrals, the so-called master integrals, using the integration by parts method [4]. We discuss all that with the help of examples in chapter 2.

In this thesis we consider mainly the virtual contributions to single top quark production in two parts, one-loop corrections squared and two-loop QCD corrections. Since each part can be considered as a gauge invariant part we treat them separately. In chapter 3 after a brief review of the properties of the single top quark production and of the motivation for the calculation of this process at NNLO, we apply the methods from chapter 2 to calculate the one-loop corrections to the single top quark production at NNLO.

To verify our setup and provide the first independent check to the calculation of the axial vector coupling [3], in chapter 4 we calculate the axial vector form factors. This consistency check ensures that our setup is correct before we begin with the most complicated part of the calculation.

In the last part of this thesis we calculate the two-loop contributions to single top quark production at NNLO. In the first part of chapter 5 we calculate the vertex contributions to single top quark production at NNLO and then in the second part in section 5.3, we discuss the contribution of the double box diagrams. This calculation requires significantly more effort compared to the vertex contributions because we have one more variable in the problem, the mass of the W boson. The main step for this part is to find master integrals through reduction with the integration by parts method. We discuss some subtle technical details which are applied to make the reduction to master integrals possible. In this section 5.3 we discuss at first how the reduction works considering an implemented Laporta algorithm [24]. Then we explain the main points and features of Reduze2 [25]. Finally, we combine the knowledge from these two parts to organize the reduction.

2

Methods for multi-loop calculations

In this chapter we discuss the technical methods which are used for the calculations in the next chapters. We do not attempt to give a review of all techniques for multi-loop calculation but give a presentation of the methods we used in our calculations. We try to elucidate through simple examples how one can apply these methods to multi-loop calculations.

2.1. From tensor integrals to scalar integrals

In multi-loop calculations one has to deal with so-called tensor integrals. We call a loop integral, in which the loop momentum occurs also in the numerator a *tensor integral*. A *scalar integral* is a loop integral, in which the numerator is independent of the loop momentum. Tensor integrals have to be reduced to scalar integrals, so one can calculate them more easily. We discuss initially the Passarino–Veltman reduction [26] which was historically the first systematic method used to reduce one-loop Feynman integrals. Then we discuss two more general methods for the application to multi-loop tensor integrals.

2.1.1. Passarino-Veltman reduction

Consider a generic one-loop diagram with n propagators

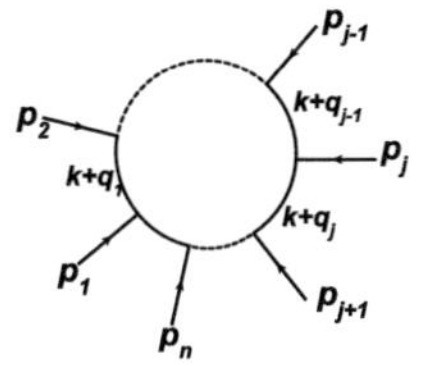

Figure 2.1.: I_n loop integral

$$I_n[f(k)] = -i(4\pi)^{d/2} \int \frac{d^d k}{(2\pi)^d} \frac{f(k)}{(k^2 - m_0^2 + i\delta)((k+q_1)^2 - m_1^2 + i\delta) \cdots ((k+q_{n-1})^2 - m_{n-1}^2 + i\delta)}, \quad (2.1)$$

where

$$q_i = p_1 + p_2 + \cdots p_i,$$

with p_i being the (incoming) external momenta. $d = 4 - 2\epsilon$ is the space-time dimension. δ is real and positive according to the Feynman prescription. $i\delta$ keeps track of the directions into which the contour has to be deformed. This term will be suppressed in order to keep the notation compact. We assume that all external momenta are in four dimensions. We also define the integrals I_{n-1}^j

$$I_{n-1}^j[f(k)] = -i(4\pi)^{d/2} \int \frac{d^d k}{(2\pi)^d} f(k)$$

$$\times \frac{1}{(k^2 - m_0^2)((k+q_1)^2 - m_1^2) \cdots ((k+q_{j-1})^2 - m_{j-1}^2)}$$

$$\times \frac{1}{((k+q_{j+1})^2 - m_{j+1}^2) \cdots ((k+q_{n-1})^2 - m_{n-1}^2)}, \quad (2.2)$$

i.e., the $(n-1)$–point integral is obtained by "pinching out" the j-th propagator. One can

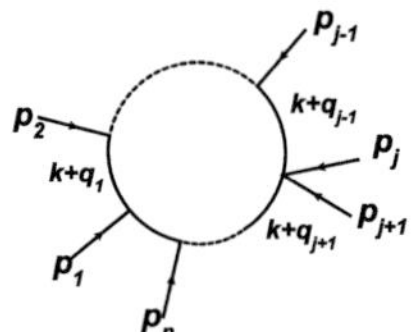

Figure 2.2.: I_{n-1} loop integral

define the integral $I_{n-2}^{j_1 j_2}$ in a similar way .

Due to Lorentz symmetry we can write the integral in eq. (2.1) in the case $f(k) = k^\mu$ as [26]

$$I_n[k^\mu] = -i(4\pi)^{d/2} \int \frac{d^d k}{(2\pi)^d} \frac{k^\mu}{(k^2 - m_0^2)((k+q_1)^2 - m_1^2) \cdots ((k+q_{n-1})^2 - m_{n-1}^2)} = \sum_{i=1}^{n-1} C_{n;i} p_i^\mu. \quad (2.3)$$

Contracting both sides with p_j^μ leads to

$$I_n[k \cdot p_j] = -i(4\pi)^{d/2} \int \frac{d^d k}{(2\pi)^d} \frac{k \cdot p_j}{(k^2 - m_0^2)((k+q_1)^2 - m_1^2) \cdots ((k+q_{n-1})^2 - m_{n-1}^2)} = \sum_{i=1}^{n-1} C_{n;i} \Delta^{ij}, \quad (2.4)$$

where $\Delta^{ij} = p_i \cdot p_j$ is the Gram matrix. Since $p_j = q_j - q_{j-1}$ (with $q_0 = 0$) one can write the numerator of the integral as

$$k \cdot p_j = \frac{1}{2}\left(((k+q_j)^2 - m_j^2) - ((k+q_{j-1})^2 - m_{j-1}^2) + m_j^2 - m_{j-1}^2 - q_j^2 + q_{j-1}^2 \right).$$

This leads to the *Passarino-Veltman* reduction formula. The terms $((k+q_j)^2 - m_j^2)$ and $((k+q_{j-1})^2 - m_{j-1}^2)$ in the numerator can be used to cancel (or "pinch") the j-th and $(j-1)$-th propagators respectively, and so we end up with a set of $n-1$ linear equations for the coefficients $C_{n;i}$

$$\sum C_{n;i}\Delta^{ij} = \frac{1}{2}\left(I_{n-1}^{(j)}[1] - I_{n-1}^{(j-1)}[1] + (m_j^2 - m_{j-1}^2 - q_j^2 + q_{j-1}^2)I_n[1] \right). \tag{2.5}$$

This set of linear equations is readily solved if Δ^{-1} exists,

$$C_{n;i} = \frac{1}{2}\sum (\Delta^{-1})^{ij}\left(I_{n-1}^{(j)}[1] - I_{n-1}^{(j-1)}[1] + (m_j^2 - m_{j-1}^2 - q_j^2 + q_{j-1}^2)I_n[1] \right). \tag{2.6}$$

However, care is needed if we are dealing with $n > 4$. The reason is that in this case only four of the external momenta p_i can be linearly independent.
In such cases, it is sufficient to choose the first four external momenta only provided that the *Gram determinant* for those four does not vanish

$$det(\Delta^{ij}) \neq 0 . \tag{2.7}$$

In special cases, known as *exceptional momenta*, this determinant may indeed vanish and we must then choose a different set of four momenta.
   This process can be repeated for any number of powers of the loop momentum in the numerator [27, 28]. However, the expressions become increasingly longer. For two powers of the loop momentum one has

$$I_n[k^\mu k^\nu] = C_{n;00}g^{\mu\nu} + \sum_{i,j} C_{n;i,j}p_i^\mu p_j^\nu. \tag{2.8}$$

For three powers of the loop momentum one has

$$I_n[k^\mu k^\nu k^\rho] = \sum_{i=1} C_{n;00i}(g^{\mu\nu}p_i^\rho + g^{\nu\rho}p_i^\mu + g^{\rho\mu}p_i^\nu) + \sum_{i,j,k} C_{n;ijk}p_i^\mu p_j^\nu p_k^\rho, \tag{2.9}$$

which can be contracted with the external momenta and $g^{\mu\nu}$ to obtain a set of linear equations for the coefficients $C_{n;00i}$ or $C_{n;ijk}$[1].
   The Passarino-Veltman algorithm is based on the observation that for one-loop integrals a scalar product of the loop momentum with an external momentum can be expressed as a combination of inverse propagators. This property does no longer hold if one goes to two or more loops. In fig. 2.3 one can see an example of a two-loop diagram, for which the scalar product of a loop momentum k_1 with an external momentum p can not be expressed in terms of inverse propagators.

---

[1]One can see for more details [26] and appendix A in [29].

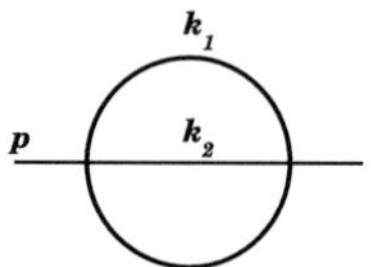

Figure 2.3.: Irreducible scalar product in the numerator

2.1.2. General reduction method

A general method to reduce tensor integrals by constructing differential operators which shift the powers of the propagators as well as the dimension of the integral was introduced by Tarasov [30, 31]. We can apply this method to higher tensor structures.

For the application of this method one may use the *Schwinger parametrization*

$$\frac{1}{Q_1^{\nu_1}\cdots Q_n^{\nu_n}} = \int \mathcal{D}x\, e^{\sum_{i=1}^{n} x_i Q_i} \tag{2.10}$$

of a tensor integral, where Q_i are the propagators and ν_i the corresponding power of the i-th propagator. The "Schwinger measure" is given by

$$\int \mathcal{D}x = \prod_{i=1}^{n} \frac{(-1)^{\nu_i}}{\Gamma(\nu_i)} \int_0^\infty dx_i x_i^{\nu_i-1}. \tag{2.11}$$

With a suitable transformation of the loop momenta we can bring the polynomial $\sum_{i=1}^{n} x_i Q_i$ in eq. (2.10) to a quadratic form in k. Integrals with an odd power of the loop momentum in the numerator vanish by symmetry, while integrals with an even power of the loop momentum can be related to scalar integrals by Lorentz invariance

$$\int \frac{d^d l}{i\pi^{d/2}} l^\mu l^\nu f(l^2) = -\frac{1}{d}g^{\mu\nu} \int \frac{d^d l}{i\pi^{d/2}} (-l^2) f(l^2). \tag{2.12}$$

In order to explain this method, we consider a simple one-loop integral with which we represent the reduction algorithm. A tensor integral is given in the following notation

$$I^d(\nu_1, \nu_2, \cdots; Q_1, Q_2, \cdots)[1, l^\mu, l^\mu l^\rho, \cdots] = \int \frac{d^d l}{i\pi^{d/2}} \frac{[1, l^\mu, l^\mu l^\rho, \cdots]}{\prod_{i=1}(Q_i^2)^{\nu_i}}, \tag{2.13}$$

where Q_i are the propagators and ν_i the corresponding power of the i-th propagator which is normally equal to one. d in I^d denotes the space-time dimension, $[1, l^\mu, l^\mu l^\rho]$ provides the information about the structure of the numerator. In a concrete example we consider the following tensor integral

$$I^d(\nu_1, \nu_2, \nu_3; Q_1, Q_2, Q_3)[l^\mu l^\rho] = \int \frac{d^d l}{i\pi^{d/2}} \frac{l^\mu l^\rho}{(l^2)^{\nu_1}((l+p_2)^2)^{\nu_2}((l+p_3)^2)^{\nu_3}}, \tag{2.14}$$

where $Q_1 = l^2$, $Q_2 = (l + p_1)^2$ and $Q_3 = (l + p_2)^2$ and p_1 and p_2 are the external momenta. We introduce the Schwinger parametrization (see eq.(2.10)) and rewrite the integrand

$$\int \frac{d^d l}{i\pi^{d/2}} \frac{l^\mu l^\rho}{(l^2)^{\nu_1}((l + p_2)^2)^{\nu_2}((l + p_3)^2)^{\nu_3}} = \int \mathcal{D}x \int \frac{d^d l}{i\pi^{d/2}} (l^\mu l^\rho) \, e^{\sum_{i=1}^3 x_i Q_i}. \tag{2.15}$$

The sum in the exponent can be written as

$$\sum_{i=1}^3 x_i Q_i = x_1 l^2 + x_2(l + p_2)^2 + x_3(l + p_3)^2$$

$$= l^2 \underbrace{(x_1 + x_2 + x_3)}_{:=a} + 2 \underbrace{(x_2 p_2 + x_3 p_3)}_{:=d} l + \underbrace{(x_2 p_2^2 + x_3 p_3^2)}_{:=r}$$

$$= a l^2 + 2\boldsymbol{d} \cdot l + r = a(l + \frac{\boldsymbol{d}}{a})^2 + \underbrace{(-\frac{\boldsymbol{d}^2}{a} + r)}_{:=R}$$

$$= a k^2 + R, \tag{2.16}$$

where we have defined[2]

$$a = x_1 + x_2 + x_3, \qquad \boldsymbol{d}^\mu = x_2 p_2^\mu + x_3 p_3^\mu, \qquad r = x_2 p_2^2 + x_3 p_3^2,$$

$$k^\mu = l^\mu + \frac{\boldsymbol{d}^\mu}{a}, \qquad R = r - \frac{\boldsymbol{d}^2}{a}. \tag{2.17}$$

By substituting $l = k - \frac{\boldsymbol{d}}{a}$ in eq. (2.15) one obtains

$$\int \mathcal{D}x \int \frac{d^d l}{i\pi^{d/2}} (l^\mu l^\rho) e^{\sum_{i=1}^3 x_i Q_i} = \int \mathcal{D}x \int \frac{d^d k}{i\pi^{d/2}} (k - \frac{\boldsymbol{d}}{a})^\mu (k - \frac{\boldsymbol{d}}{a})^\rho e^{(ak^2 + R)} =$$

$$\int \mathcal{D}x \int \frac{d^d k}{i\pi^{d/2}} \left(k^\mu k^\rho + (-\frac{\boldsymbol{d}^\mu}{a} k^\rho - \frac{\boldsymbol{d}^\rho}{a} k^\mu) + \frac{\boldsymbol{d}^\mu \boldsymbol{d}^\rho}{a^2} \right) e^{(ak^2 + R)} =$$

$$\underbrace{\int \mathcal{D}x \int \frac{d^d k}{i\pi^{d/2}} k^\mu k^\rho e^{(ak^2 + R)}}_{\text{(I)}} - \underbrace{\int \mathcal{D}x \int \frac{d^d k}{i\pi^{d/2}} (\frac{\boldsymbol{d}^\mu}{a} k^\rho + \frac{\boldsymbol{d}^\rho}{a} k^\mu) e^{(ak^2 + R)}}_{\text{(II)}} +$$

$$\underbrace{\int \mathcal{D}x \int \frac{d^d k}{i\pi^{d/2}} \frac{\boldsymbol{d}^\mu \boldsymbol{d}^\rho}{a^2} e^{(ak^2 + R)}}_{\text{(III)}}. \tag{2.18}$$

Each of the three integrals in eq.(2.18) will be evaluated separately. We begin with the easiest integral, the second one (II) . It vanishes because we have an odd function of k

---

[2]We write d bold because we want to distinguish between d, space-time dimension, and the vector $\boldsymbol{d}$.

$$\int \frac{d^d k}{i\pi^{d/2}} \left(\frac{\boldsymbol{d}^{\mu}}{a} k^{\rho} + \frac{\boldsymbol{d}^{\rho}}{a} k^{\mu}\right) e^{(ak^2+R)} = 0.$$ (2.19)

The first integral ① is

$$\int \frac{d^d k}{i\pi^{d/2}} k^{\mu} k^{\rho} e^{(ak^2+R)} = -\frac{1}{2a} g^{\mu\rho} \frac{1}{a^{d/2}} e^{R},$$ (2.20)

since

$$\int \frac{d^d k}{i\pi^{d/2}} k^{\mu} k^{\rho} f(k^2) \propto g^{\mu\rho}.$$ (2.21)

When we multiply both sides of eq.(2.21) with $g^{\mu\rho}$ and perform the Gaussian integral we get

$$\int \frac{d^d k}{i\pi^{d/2}} k^2 e^{ak^2} = -\frac{d}{2}\frac{1}{a}\left(\frac{1}{a}\right)^{d/2}.$$ (2.22)

As one can see from eq.(2.22), $\frac{1}{a}$ shifts the dimension of the integral

$$\int \frac{d^{d+2} k}{i\pi^{(d+2)/2}} k^2 e^{ak^2} = -\frac{d+2}{2}\frac{1}{a}\left(\frac{1}{a}\right)^{d/2}\frac{1}{a}.$$ (2.23)

In order to calculate the third integral ⑪, one changes the order of integrations

$$\int \mathcal{D}x \int \frac{d^d k}{i\pi^{d/2}} \frac{\boldsymbol{d}^{\mu} \boldsymbol{d}^{\rho}}{a^2} e^{(ak^2+R)} = \int \frac{d^d k}{i\pi^{d/2}} \int \mathcal{D}x \frac{\boldsymbol{d}^{\mu} \boldsymbol{d}^{\rho}}{a^2} e^{(ak^2+R)}$$

$$= \int \frac{d^d k}{i\pi^{d/2}} \prod_{i=1}^{3} \frac{(-1)^{\nu_i}}{\Gamma(\nu_i)} \int dx_1 dx_2 dx_3\, x_1^{\nu_1-1} x_2^{\nu_2-1} x_3^{\nu_3-1} \frac{\boldsymbol{d}^{\mu} \boldsymbol{d}^{\rho}}{a^2} e^{(ak^2+R)}.$$ (2.24)

By the substitution of $\boldsymbol{d}^{\mu}$ from eq.(2.17), one gets the following integral

$$\int \frac{d^d k}{i\pi^{d/2}} \prod_{i=1}^{3} \frac{(-1)^{\nu_i}}{\Gamma(\nu_i)} \int dx_1 dx_2 dx_3\, x_1^{\nu_1-1} x_2^{\nu_2-1} x_3^{\nu_3-1} \frac{\boldsymbol{d}^{\mu} \boldsymbol{d}^{\rho}}{a^2} e^{(ak^2+R)} = \int \frac{d^d k}{i\pi^{d/2}} \prod_{i=1}^{3} \frac{(-1)^{\nu_i}}{\Gamma(\nu_i)} \times$$

$$\int dx_1 dx_2 dx_3\, x_1^{\nu_1-1} x_2^{\nu_2-1} x_3^{\nu_3-1} \frac{(x_2 p_2 + x_3 p_3)^{\mu}(x_2 p_2 + x_3 p_3)^{\rho}}{a^2} e^{(ak^2+R)} = \int \frac{d^d k}{i\pi^{d/2}} \prod_{i=1}^{3} \frac{(-1)^{\nu_i}}{\Gamma(\nu_i)} \times$$

$$\int dx_1 dx_2 dx_3\, x_1^{\nu_1-1} x_2^{\nu_2-1} x_3^{\nu_3-1} \frac{x_2^2 p_2^{\mu} p_2^{\rho} + x_2 x_3 p_2^{\mu} p_3^{\rho} + x_3 x_2 p_3^{\mu} p_2^{\rho} + x_3^2 p_3^{\mu} p_3^{\rho}}{a^2} e^{(ak^2+R)}.$$ (2.25)

The Schwinger coefficients x_2 and x_3 will be absorbed in the power of the corresponding propagator, i.e., x_i increases the power of the i-th propagator

$$\frac{(-1)^{\nu_i}x_i^{\nu_i-1}}{\Gamma(\nu_i)}\,x_i = -\nu_i\frac{(-1)^{\nu_i+1}x_i^{\nu_i}}{\Gamma(\nu_i+1)} := -\nu_i\mathbf{i}^+, \tag{2.26}$$

where $\mathbf{i}^+$ increases the power of i-th propagator.

a^2 in the denominator shifts the dimension of the integral by 4 (see eq.(2.23)), then the Schwinger parameters will be reversed, i.e., the third integral ⑪ is equal to

$$\int\frac{d^dk}{i\pi^{d/2}}\prod_{i=1}^{3}\frac{(-1)^{\nu_i}}{\Gamma(\nu_i)}\int dx_1dx_2dx_3\,x_1^{\nu_1-1}x_2^{\nu_2-1}x_3^{\nu_3-1}\frac{d^\mu d^\rho}{a^2}e^{(ak^2+R)} =$$

$$p_2^\mu p_2^\rho\int\frac{d^{d+4}l}{i\pi^{(d+4)/2}}\frac{(\nu_2+1)\nu_2}{(l^2)^{\nu_1}((l+p_1)^2)^{\nu_2+2}((l+p_2)^2)^{\nu_3}}+$$

$$p_2^\mu p_3^\rho\int\frac{d^{d+4}l}{i\pi^{(d+4)/2}}\frac{\nu_2\nu_3}{(l^2)^{\nu_1}((l+p_1)^2)^{\nu_2+1}((l+p_2)^2)^{\nu_3+1}}+$$

$$p_3^\mu p_2^\rho\int\frac{d^{d+4}l}{i\pi^{(d+4)/2}}\frac{\nu_2\nu_3}{(l^2)^{\nu_1}((l+p_1)^2)^{\nu_2+1}((l+p_2)^2)^{\nu_3+1}}+$$

$$p_3^\mu p_3^\rho\int\frac{d^{d+4}l}{i\pi^{(d+4)/2}}\frac{(\nu_3+1)\nu_3}{(l^2)^{\nu_1}((l+p_1)^2)^{\nu_2}((l+p_2)^2)^{\nu_3+2}}. \tag{2.27}$$

With the help of this algorithm one can write a tensor integral as a sum of scalar integrals which have propagators with different powers. With the notation of eq. (2.14) we can write

$$I^d(\nu_1,\nu_2,\nu_3;Q_1,Q_2,Q_3)[l^\mu l^\rho] = -\frac{g^{\mu\rho}}{2}I^{d+2}(\nu_1,\nu_2,\nu_3;Q_1,Q_2,Q_3)[1]+$$

$$(\nu_2+1)\nu_2 p_2^\mu p_2^\rho I^{d+4}(\nu_1,\nu_2+2,\nu_3;Q_1,Q_2,Q_3)[1] + \nu_2\nu_3 p_2^\mu p_3^\rho I^{d+4}(\nu_1,\nu_2+1,\nu_3+1;Q_1,Q_2,Q_3)[1]+$$

$$\nu_2\nu_3 p_3^\mu p_2^\rho I^{d+4}(\nu_1,\nu_2+1,\nu_3+1;Q_1,Q_2,Q_3)[1] + (\nu_3+1)\nu_3 p_3^\mu p_3^\rho I^{d+4}(\nu_1,\nu_2,\nu_3+2;Q_1,Q_2,Q_3)[1]. \tag{2.28}$$

As we can see after reduction of a given tensor integral one obtains many scalar integrals with the same structure of the propagators, but with different powers and a shifted dimension. We name the increase of the power of each propagator by one a *dot*. This means, the i-th propagator with the power ν_i+1, $(\nu_i=1)$ has a *dot*, and the j-th propagator with the power ν_j has (ν_j-1) *dots*.

Now we apply this method to two-loop tensor integrals. We diagonalize the argument of the exponential function after the Schwinger parametrization of an integral. Integrals with an odd power of the loop momentum in the numerator vanish by symmetry, while integrals with an even power of the loop momentum can be expressed as a sum of scalar integrals with different powers of the propagators and a shifted dimension of the scalar integrals.

A two-loop tensor integral has the following Schwinger parametrization

$$\int \mathcal{D}x\int\frac{d^dl_1}{i\pi^{d/2}}\int\frac{d^dl_2}{i\pi^{d/2}}[l_1^\mu,l_2^\mu,l_1^\mu l_1^\rho,l_2^\mu l_2^\rho,l_1^\mu l_2^\rho,\cdots]\,e^{\sum_{i=1}^{n}x_iQ_i}, \tag{2.29}$$

where Q_i are the propagators and $\mathcal{D}x$ is the Schwinger measure (see eq. (2.11)).

In this case one has two loop momenta, l_1 and l_2. The argument of the exponential function can be written as

$$\sum_{i=1}^{n} x_i Q_i = al_1^2 + bl_2^2 + 2cl_1l_2 + 2\boldsymbol{d} \cdot l_1 + 2\boldsymbol{e} \cdot l_2 + f. \tag{2.30}$$

The coefficients $a, b, c, \boldsymbol{d}, \boldsymbol{e}$ and f depend on the topology of the integral [32] and are different for various topologies. After the determination of the coefficients, one can make the following substitutions

$$l_1 = k_1 - \frac{c}{a}k_2 + X \qquad l_2 = k_2 + Y, \tag{2.31}$$

where

$$X = \frac{ce - bd}{P}, \qquad Y = \frac{cd - ae}{P} \tag{2.32}$$

and

$$P = ab - c^2. \tag{2.33}$$

The sum in eq.(2.30) can now be diagonalized and written as

$$\sum_{i=1}^{n} x_i Q_i = ak_1^2 + \frac{P}{a}k_2^2 + \frac{Q}{P}, \tag{2.34}$$

where

$$Q = -ae^2 - bd^2 + 2ce \cdot d + fP. \tag{2.35}$$

After the parametrization, i.e., the application of eq. (2.31) and eq. (2.32), one has either integrals with an odd power of the loop momentum in the numerator

$$\int \frac{d^d k}{i\pi^{d/2}} k^\mu \exp(ak^2) = 0, \tag{2.36}$$

or integrals with an even power of the loop momentum

$$\int \frac{d^d k}{i\pi^{d/2}} k^\mu k^\nu \exp(ak^2) = -\frac{1}{2a} g^{\mu\nu} \frac{1}{a^{D/2}}$$

$$\int \frac{d^d k}{i\pi^{d/2}} k^\mu k^\nu k^\rho k^\sigma \exp(ak^2) = \frac{1}{4a^2} \{g^{\mu\nu} g^{\rho\sigma} + g^{\mu\rho} g^{\nu\sigma} + g^{\mu\sigma} g^{\nu\rho}\} \frac{1}{a^{D/2}}. \tag{2.37}$$

Just as in the one-loop case, the i-th Schwinger coefficient x_i increases the power of the i-th propagator. Further, $1/P$ acts like $1/a^2$ in the one-loop case and shifts the dimension of the integral, as one can see in eq.(2.25) and eq.(2.27)

$$\frac{1}{P} \to d^+, \ P \to d^-,$$

(2.38)

i.e.,

$$d^\pm I^d(\{v_i\}; \{Q_i^2\}) = I^{d\pm2}(\{v_i\}; \{Q_i^2\}).$$

(2.39)

It should be noted that both integration measures, dk_1 and dk_2, are shifted at the same time in the space-time dimension.

Using the method described above each tensor integral can be reduced to a sum of scalar integrals with the same structure of the propagators, but with different powers and shifted dimensions. However, depending on the rank of the tensor integral and the number of loops, we obtain various numbers of **dots** in the scalar integrals, as one can see in eq.(2.28). There is another method which does not produce any **dots** but we get scalar products between loop momenta and external momenta which can be expressed by negative powers of the propagators. We will now continue with the so-called "projection method" [33–35], which we will use for tensor reduction, too.

2.1.3. Projection method

In our case, except for the colour coefficient before the tensor reduction an amplitude can be written as

$$\mathcal{A} = \sum_i \overline{u}(p_1)f_i(\gamma, \{\not k\}, \{\not p\})u(p_2)\overline{u}(p_3)g_i(\gamma, \{\not k\}, \{\not p\})u(p_4),$$

(2.40)

where $f_i(\gamma, \{\not k\}, \{\not p\})$ and $g_i(\gamma, \{\not k\}, \{\not p\})$ are functions depending on the gamma matrices γ, the loop momenta $\{\not k\}$ and the external momenta $\{\not p\}$. $u(p_i)(i = 1, \cdots, 4)$ are Dirac spinors with momenta p_i.

In order to reduce the tensor structure, in other words to eliminate loop momenta in eq.(2.40), one may find at first projectors as follows

$$\mathcal{S} = \sum_j \overline{u}(p_1)h_j(\gamma, \{\not p\})u(p_2)\overline{u}(p_3)r_j(\gamma, \{\not p\})u(p_4),$$

(2.41)

where $h_i(\gamma, \{\not p\})$ and $r_i(\gamma, \{\not p\})$ are functions depending on the gamma matrices γ and the external momenta $\{\not p\}$. In the next step we multiply the amplitude in eq. (2.40) with the projectors in eq. (2.41) and take the sum over the spins

$$\sum_{spins} \mathcal{A} \mathcal{S}^\dagger.$$

(2.42)

On the other hand, the most general tensor structure for the amplitude $\mathcal{A}$ can be written as

$$A = \sum_{i=1}^{n} B_i(t, u, s)\mathcal{S}_i,\tag{2.43}$$

where t, u and s are the *Mandelstam variables* and $\mathcal{S}_i$ are the *Dirac structures*. If we know the Dirac structures $\mathcal{S}_i$ and the coefficients B_i we can construct the amplitude A. However, we don't know the coefficients B_i. In order to determine them one shall know the projectors. For this aim we use the occurring spin structures. One can determine them in different ways. We used the previous method, described in subsec. 2.1.2, to eliminate the loop momenta and to get expressions containing spinors, gamma matrices and external momenta. To find the coefficients B_i we multiply eq. (2.43) with $\mathcal{S}_j^\dagger$

$$\mathcal{S}_j^\dagger A = \sum_{i=1}^{n} B_i(t, u, s)\underbrace{(\mathcal{S}_j^\dagger \mathcal{S}_i)}_{\mathcal{M}_{ji}}\tag{2.44}$$

and by solving eq. (2.44) according to $B_i(t, u, s)$ we find

$$B_i(t, u, s) = \sum_j \mathcal{M}_{ij}^{-1}(\mathcal{S}_j^\dagger A).\tag{2.45}$$

Considering eq.(2.43) and eq.(2.45) we can now obtain the amplitude A occurring in eq.(2.42)

$$A = \sum_{i=1}^{n} \mathcal{M}_{ij}^{-1}(\mathcal{S}_j^\dagger A)\mathcal{S}_i.\tag{2.46}$$

We now want to illustrate this method by using an example. The amplitude of the

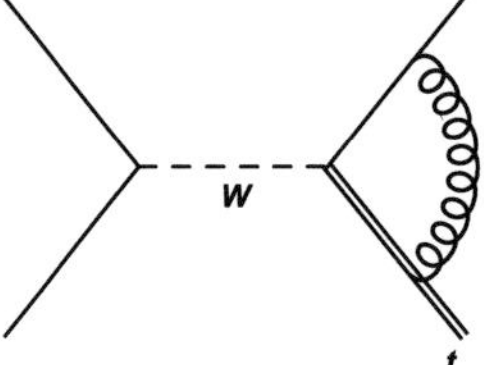

Figure 2.4.: Projection method

process in fig. 2.4, except for a few factors is given by

$$A = \int \frac{d^d k}{(2\pi)^d} \frac{1}{(k^2)\,(p_2 - k)^2\,((q_1 - k)^2 - m_t^2)} \times$$
$$\left(\bar{u}(q_2)\gamma_\nu\gamma_7 u(p_1)\right) \times \left(\bar{u}(q_1)\gamma_\mu(\not{q}_1 - \not{k} + m_t)\gamma_\nu\gamma_7(\not{p}_2 - \not{k})\gamma_\mu u(p_2)\right).\tag{2.47}$$

The following spin structures occur in the diagram in fig. 2.4

$$S_1 = \bar{u}(q_1)\gamma_7 u(p_2)\,\bar{u}(q_2)\gamma_6 \slashed{q}_1 u(p_1),$$
$$S_2 = \bar{u}(q_1)\gamma_6\gamma_\mu u(p_2)\,\bar{u}(q_2)\gamma_6\gamma_\mu u(p_1), \tag{2.48}$$

where γ_μ is the gamma matrix and $\gamma_6 = 1 + \gamma_5$ and $\gamma_7 = 1 - \gamma_5$. $u(p_i)$ and $u(q_i)(i = 1, 2)$ are Dirac spinors with momenta p_i incoming and q_i outgoing. q_1 is the momentum of the top quark. In order to calculate the coefficients B_i, we build in the first step the product $S_j^\dagger S_i$

$$\begin{pmatrix} S_1 \\ S_2 \end{pmatrix}^\dagger \otimes \begin{pmatrix} S_1 \\ S_2 \end{pmatrix} \tag{2.49}$$

and calculate the trace of the gamma matrices, then we obtain

$$\mathcal{M}_{ji} = \begin{pmatrix} -4st(s+t) & -8m_t st \\ -8m_t st & 4\big((-2+d)s^2 + 2(-4+d)st + (-2+d)t^2 - (-2+d)m_t^2(s+t)\big) \end{pmatrix}, \tag{2.50}$$

where t and s are the Mandelstam variables

$$p_1 \cdot p_2 = \frac{s}{2}, \quad p_2 \cdot q_2 = -\frac{t}{2}, \tag{2.51}$$

m_t the top mass and d the dimension.

The inverse of this matrix should be calculated and then multiplied with (S_1, S_2) as in eq. (2.45). Therefore, we get for B_i $(i = 1, 2)$

$$B_1 = \frac{-2m_t st S_2 + (-2+d)m_t^2 S_1(s+t) - S_1((-2+d)s^2 + 2(-4+d)st + (-2+d)t^2)}{4st(-m_t^2 + s + t)((-2+d)s^2 + 2(-4+d)st + (-2+d)t^2)}, \tag{2.52}$$

$$B_2 = \frac{2m_t S_1 - S_2(s+t)}{4(m_t^2 - s - t)((-2+d)s^2 + 2(-4+d)st + (-2+d)t^2)}. \tag{2.53}$$

Multiplying the amplitude $\mathcal{A}$ in eq.(2.47) with the conjugated spin structures occurring in eq.(2.48) and summing over the spins as in eq.(2.46), leads to

$$\begin{aligned}
\mathcal{A} = A_m \times \Big[&-8m_t B_1 \Big(2st(s+t) + (d-2)st\, k \cdot k - tp_1 \cdot k(3s + t + 2(d-2)p_2 \cdot k) \\
&+ (m_t^2 - s - t)((s+t)p_2 \cdot k + 2(d-2)p_2 \cdot k^2) + s(s + 3t + 2(d-2)p_2 \cdot k)q_2 \cdot k \Big) \\
&- 4B_2 \Big((d-2)(-(d-4)s^2 - 2(d-6)st - (d-4)t^2 + (d-4)m_t^2(s+t))k \cdot k \\
&+ 2((s+t)(-(d-2)s^2 - 2(d-4)st - (d-2)t^2 + (d-2)m_t^2(s+t)) - 2(d-2)t(p_1 \cdot k)^2 \\
&+ 2p_1 \cdot k((d-2)(m_t^2 - s - 3t)p_2 \cdot k + (s+t)(-(d-2)(m_t^2 - s) + (d-4)t - (d-2)q_2 \cdot k)) \\
&+ 2(2(d-2)(m_t^2 - s - t)(p_2 \cdot k)^2 + (s+t)((d-2)m_t^2 - d(s+t) + 2(2s+t))q_2 \cdot k
\end{aligned}$$

$$+ p_2 \cdot k((m_t^2 - s - t)((d-2)m_t^2 - 2(d-3)(s+t)) - (d-2)(m_t^2 - 3s - t)q_2 \cdot k)$$
$$- (d-2)s(q_2 \cdot k)^2)) \Big], \tag{2.54}$$

where $A_m := \int \frac{d^d k}{(2\pi)^d} \frac{1}{(k^2)\,(p_2-k)^2\,((q_1-k)^2-m_t^2)}$, B_1 and B_2 are the coefficients in eq.(2.52) and eq.(2.53) and $k \cdot p_i$ and $k \cdot q_i$ ($i = 1,2$) are scalar products. We can express these scalar products by negative powers of the propagators. As we can see, by using this method we get no shift in the dimension and no dots (see eq.(2.28)).

2.2. From scalar integrals to master integrals

In the previous section, we saw that the tensor reduction usually produces a huge number of scalar integrals which have the same structure of the integrand with various powers of propagators and partly different dimensions

$$I(a_1, \cdots, a_n) = \int \cdots \int \frac{d^d k_1 \cdots d^d k_h}{Q_1^{a_1} \cdots Q_n^{a_n}}, \tag{2.55}$$

where $d = 4 - 2\epsilon$ and $Q_i, i = 1, \cdots, n$ are the propagators. Note, that the integral on the left hand side still depends on other variables like the external momenta. However, for the explanation in the following the dependency on the powers is essential. We can assume in general that $a_i \neq 1$ for $i = 1, \cdots, n$. There are different methods to evaluate these integrals. We can either evaluate every scalar Feynman integral individually or we can look for an alternative method. The method of integration by parts (IBP) principally enables us to express all occurring scalar integrals after tensor reduction as a linear combination of some basic *master integrals*. Thus, the whole problem of evaluation is divided into two steps: constructing a reduction procedure to find some master integrals, which enable us to express each scalar integral as a linear combination of them, and evaluating these master integrals.

In case of the shifted dimension (see eq.(2.28)) we have to shift the dimension back to d. This step will be explained in sec. 2.3.2.

To illustrate the procedure of solving IBP relations, we begin with an introduction and will then discuss some simple examples.

2.2.1. Integration by parts

The basic idea of the integration by parts method is to find relations between scalar integrals with various powers of the propagators. The most common relation [4] which is used to find IBP identities is

$$\int d^d k_1 \cdots d^d k_h \frac{\partial}{\partial k^\mu} v^\mu \frac{1}{Q_1^{a_1} \cdots Q_n^{a_n}} = 0, \tag{2.56}$$

where k is a loop momentum and v is either an external momentum or a loop momentum. After the application of the IBP identity, eq.(2.56), we obtain some relations of the following form

$$\sum_i \alpha_i I(a_1 + b_{i,1}, \cdots, a_n + b_{i,n}) = 0, \tag{2.57}$$

where $b_{i,j}, j = 1, \cdots, n$ are integers and α_i are inter alia polynomials in a_i.

There are some more relations which can be used: Lorentz-invariance (*LI*) identities [36], and symmetry relations

$$I(a_1, \cdots, a_n) = I(a_{\sigma(1)}, \cdots, a_{\sigma(n)}), \tag{2.58}$$

where σ is a permutation.

All these relations in eq.(2.57) enable us to express a given Feynman integral in terms of other Feynman integrals. Therefore, we have to find some irreducible integrals, named master integrals, and aim to reduce any other integral to those. There are some attempts to define master integrals [37]. For instance, we can choose integrals with more non-positive indices as master integrals.

One can solve the relations in eq.(2.57) by hand, but as soon as the complexity of the problem grows, it turns out to be impractical and unfeasible. The first algorithmic approach to solve IBP relations was the so-called **Laporta's** algorithm [24] which is actively used in practice.

There are other algorithmic approaches to solve the relations such as the Baikov's method [38–41] and the use of Gröbner bases [42].

We will illustrate the method of integration by parts by using some examples.

2.2.2. Some examples to IBP

We consider the massive one-loop Feynman integral

$$I(a) = \int \frac{d^d k}{(k^2 - m^2)^a}. \tag{2.59}$$

The dependence of the integral on the power of the propagators is indicated explicitly as it plays a key role in the following discussion. We actually know the result of the integral in eq.(2.59). However, we illustrate the application of IBP method to this integral: We use the following identity

$$\int d^d k \frac{\partial}{\partial k} \cdot k \frac{1}{(k^2 - m^2)^a} = 0. \tag{2.60}$$

At first we have to calculate the derivative

$$\frac{\partial}{\partial k} \cdot k \frac{1}{(k^2 - m^2)^a} = \frac{d}{(k^2 - m^2)^a} + k \cdot \frac{-a(k^2 - m^2)^{a-1} 2k}{(k^2 - m^2)^{2a}}$$

$$= \frac{d}{(k^2 - m^2)^a} - \frac{2a}{(k^2 - m^2)^a} - \frac{2am^2}{(k^2 - m^2)^{a+1}}. \tag{2.61}$$

Then we obtain the following equation

$$(d - 2a)I(a) - 2am^2 I(a + 1) = 0. \tag{2.62}$$

This equation gives us the following recursive relation between scalar integrals with different powers of the propagators

$$I(a) = \frac{d - 2a + 2}{2(a - 1)m^2} I(a - 1). \tag{2.63}$$

Eq. (2.63) enables us to express each integral with a power $a > 1$ recursively by the integral $I(a = 1) = I(1)$. The integral $I(1)$ is called the master integral.

This was a simple example in which there was only one propagator. However, usually several propagators appear with different powers in the denominator, which makes the reduction more difficult. In the next example we consider a massless one-loop integral

$$I(a_1, a_2) = \int d^d k \frac{d^d k}{(k^2)^{a_1} [(q - k)^2]^{a_2}}, \tag{2.64}$$

where q is the external momentum, and we do not consider the q- and d-dependence of the integral. We actually know the explicit result of this integral but we are applying the IBP relation to reduce the calculation of this integral to the calculation of some master integrals

$$\int d^d k \frac{\partial}{\partial k} \cdot k \frac{1}{(k^2)^{a_1} [(q - k)^2]^{a_2}} = 0, \tag{2.65}$$

i.e.,

$$\frac{\partial}{\partial k} \cdot k \frac{1}{(k^2)^{a_1} [(q - k)^2]^{a_2}} = \frac{d}{(k^2)^{a_1} [(q - k)^2]^{a_2}} + \frac{-2a_1}{(k^2)^{a_1} [(q - k)^2]^{a_2}} - \frac{a_2}{(k^2)^{a_1} [(q - k)^2]^{a_2}}$$
$$+ \frac{a_2 q^2}{(k^2)^{a_1} [(q - k)^2]^{a_2+1}} - \frac{a_2}{(k^2)^{a_1-1} [(q - k)^2]^{a_2+1}}. \tag{2.66}$$

Using the representation in eq.(2.64), we get for the IBP relation in eq.(2.65)

$$d \cdot I(a_1, a_2) - 2a_1 I(a_1, a_2) - a_2 I(a_1, a_2)$$
$$+ a_2 q^2 I(a_1, a_2 + 1) - a_2 I(a_1 - 1, a_2 + 1) = 0. \tag{2.67}$$

The change in the powers of a_1 and a_2 in $I(a_1, a_2)$ can be shown by *increasing* and *lowering* operators, i.e., $I(a_1 - 1, a_2 + 1) = 2^+ 1^- I(a_1, a_2)$ and we can rewrite eq. (2.67) as

$$(d - 2a_1 - a_2)I(a_1, a_2) - a_2 2^+ 1^- I(a_1, a_2) + a_2 q^2 2^+ I(a_1, a_2) = 0$$
$$\Rightarrow$$
$$a_2 q^2 2^+ I(a_1, a_2) = -(d - 2a_1 - a_2)I(a_1, a_2) - a_2 2^+ 1^- I(a_1, a_2). \tag{2.68}$$

Setting $a_2 = a_2 - 1$, one obtains

$$(a_2 - 1)q^2 I(a_1, a_2) = -(d - 2a_1 - a_2 + 1)2^- I(a_1, a_2) - (a_2 - 1)1^- I(a_1, a_2)$$

$$\Rightarrow$$

$$I(a_1, a_2) = \frac{-(d - 2a_1 - a_2 + 1)}{(a_2 - 1)q^2} 2^- I(a_1, a_2) - \frac{1}{q^2} 1^- I(a_1, a_2). \tag{2.69}$$

Using the symmetry property $I(a_1, a_2) = I(a_2, a_1)$, one can rewrite eq.(2.69) exchanging a_1 by a_2

$$I(a_1, a_2) = \frac{-(d - 2a_2 - a_1 + 1)}{(a_1 - 1)q^2} 1^- I(a_1, a_2) - \frac{1}{q^2} 2^- I(a_1, a_2). \tag{2.70}$$

We set $a_2 = 1$ and obtain the following equation for a_1

$$I(a_1, 1) = -\frac{d - a_1 - 1}{(a_1 - 1)q^2} I(a_1 - 1, 1). \tag{2.71}$$

This is a recursive formula that can reduce the index of a_1 to 1. With the two formulas (2.69) and (2.71) we can reduce each given integral with any a_1 and a_2 to $I(1, 1)$. The symmetry property in $I(a_1, a_2)$ is used.

We can find many other examples in [43]. It is important to note that the difficulty of finding such recursive formulas like eq. (2.69) and eq. (2.71) is increasing with the complexity of the topology and the increasing number of the propagators. One can attempt to systematize the method of solving this system of equations obtained by the IBP equations. For more relations one can use the methods based on shifting dimension [5, 31]. We can find more tricks which we can use to solve the system of equations in [30, 44].

There is another attempt of a systematization introduced by Laporta [24, 45]. It is based on the fact that the total number of IBP equations written for concrete indices grows faster than the number of independent Feynman integrals. Hence, this system of equations becomes overdetermined, and one obtains a possibility to perform a reduction to master integrals. An explicit recipe for solving overdetermined systems of equations is presented in Laporta's work [24], a Gauss-elimination-like algorithm. To apply the Laporta's algorithm one needs an ordering to make possible that more difficult integrals are expressed by simpler ones. Then the Gauss elimination is applied there. Different terms of the equations are characterized by a relative weight of their complexity, and the equations are solved starting with the most complicated terms. There are some implementations of this algorithm, the so-called Laporta's algorithm, in several languages Air [46] in Maple, Fire [47] in Mathematica, Reduze1 [48] and Reduze2 [25] in $C++$. The advantage of Reduze1&2 is that their usage of the main memory is moderate as well as that they can run parallelized, especially the full parallelization in Reduze2.

2.3. Feynman Graph Polynomial

The Feynman Graph Polynomials are interesting both from the phenomenological point of view and from the mathematical point of view. However, we discuss them here for more practical purposes in the multi-loop calculation.

In subsec. 2.1.2 we saw that all tensor integrals can be reduced to scalar integrals. The propagators of these integrals have in general a power different from one and the dimension is shifted. After obtaining all master integrals, we have to shift the dimension back to d. For this aim we need Feynman Graph Polynomials.

In the multi-loop calculation we sometimes have to calculate several hundred Feynman diagrams. But many integrals may belong to a given family ("topology"). The determination of all occurring topologies can help us to reduce the number of integrals which should be computed. Normally, in order to say whether two integrals are the same, both should be calculated. However, there is a practical way to compare two integrals, namely the comparison between graphs. Feynman Graph Polynomials enable us to compare two integrals without an explicit calculation of both of them[49]. That is important, for instance when we want to apply the IBP method to all diagrams. The first step is to find all topologies. Then we can apply this method just to the topologies and use the result for all other integrals by a suitable shift in the propagators.

There are two possible approaches to calculate the Feynman Graph Polynomials. The first one is through graph theory where one has to introduce the concepts of "spanning trees" and "spanning forests". The other one is the algebraic way. We use the algebraic way for the implementation of a program which calculates the Feynman Graph Polynomials [49, 50].

2.3.1. Identification of Feynman integrals

To a l–loop Feynman graph G with n internal edges, we associate a scalar l–loop integral I_G in d–dimensions with n propagators

$$I_G = \int \prod_{r=1}^{l} d^d k_r \prod_{j=1}^{n} \frac{1}{(-q_j^2 + m_j^2)^{a_j}}, \tag{2.72}$$

where the momenta q_i of the propagators are linear combinations of the external and the loop momenta. The Feynman propagators are allowed to appear to arbitrary integer powers a_j whose sum is denoted by $a = \sum_{j=1}^{n} a_j$. To perform the integration over the loop momenta, we convert the products of the propagators into a sum by using the Schwinger parametrization

$$\prod_{i=1}^{n} \frac{1}{(-Q_i)^{a_i}} = \prod_{i=1}^{n} \frac{1}{\Gamma(a_i)} \int_0^{\infty} dx_i x_i^{a_i-1} \exp(\sum_{i=1}^{n} x_i Q_i). \tag{2.73}$$

We apply eq.(2.73) for $Q_j = -q_j^2 + m_j^2$ to eq. (2.72). After performing the momentum integration [51] one gets the following Schwinger parameter integral [3] [31, 52, 53]

$$I_G^d = i^l \left(\frac{\pi}{i}\right)^{d\frac{l}{2}} \prod_{j=1}^{n} \frac{i^{a_j}}{\Gamma(a_j)} \int_0^\infty \frac{dx_j x_j^{a_j-1}}{[\mathcal{U}(x)]^{\frac{d}{2}}} e^{i[\frac{-\mathcal{F}}{\mathcal{U}}]}. \tag{2.74}$$

The functions $\mathcal{U}$ and $\mathcal{F}$ are called graph polynomials. $\mathcal{U}$ is called the first Symanzik polynomial and $\mathcal{F}$ the second Symanzik polynomial. These polynomials depend on the Feynman parameters x_j. We can express

$$\sum_{j=1}^{n} x_j(-q_j^2 + m_j^2) = -\sum_{r=1}^{l}\sum_{s=1}^{l} k_r M_{rs} k_s + \sum_{r=1}^{l} 2k_r \cdot Q_r + J, \tag{2.75}$$

where M is a $l \times l$ matrix with scalar entries and Q is a l-vector with four-vectors as entries. Thus, one obtains

$$\mathcal{U} = \det(M), \qquad \mathcal{F} = \det(M)\left(J + QM^{-1}Q\right). \tag{2.76}$$

The functions $\mathcal{U}$ and $\mathcal{F}$ can be straightforwardly derived from the topology of the corresponding Feynman graph G [49]. The function $\mathcal{F}$ can be rewritten in the following way

$$\mathcal{F} = \mathcal{F}_0 + \mathcal{U} \sum_{l=1}^{n} x_i m_i^2, \tag{2.77}$$

where

$$\mathcal{F}_0 = \mathcal{F}(m_i = 0). \tag{2.78}$$

This notation, eq.(2.77), helps us to write eq.(2.74) as

$$I_G^d = i^l \left(\frac{\pi}{i}\right)^{d\frac{l}{2}} \prod_{j=1}^{n} \frac{i^{a_j}}{\Gamma(a_j)} \int_0^\infty \cdots \int_0^\infty \frac{dx_j x_j^{a_j-1}}{[\mathcal{U}(x)]^{\frac{d}{2}}} e^{i[\frac{-\mathcal{F}_0}{\mathcal{U}} - \sum_{i=1}^{n} x_i m_i^2]}, \tag{2.79}$$

and it is practical for the next part, subsec. 2.3.2 . We are not going to discuss here the mathematical properties of these functions. They will be used to find relations between integrals with different dimensions as well as to find all families of topologies. We used the algorithm described in [32] to calculate the $\mathcal{U}$ polynomial which we actually need for the calculation of the shift in the dimension. But if we want to determine all topologies, we need both polynomials, $\mathcal{U}$ and $\mathcal{F}$. So we used the algorithm described in [50] to compute them.

---

[3]We assume that all a_j are integers.

2.3.2. Integrals with different dimensions

We use the parametric representation of an integral in d dimensional space-time, eq.(2.79), to find relations between integrals with different dimensions with an assumption, namely that all propagators have different masses [31]. These masses could be fictive and just help us to introduce the *polynomial differential operator*

$$\mathcal{U}(\partial) := \mathcal{U}(\frac{\partial}{\partial m_j^2}) \tag{2.80}$$

which is obtained from $\mathcal{U}(x)$ by substituting $x_j \rightarrow \partial/\partial m_j^2$. When we apply $\mathcal{U}(\partial)$ to the integral in eq. (2.79), we get $\mathcal{U}(x)$ in the numerator of the integrand

$$i^l \left(\frac{\pi}{i}\right)^{d\frac{l}{2}} \prod_{j=1}^{n} \frac{i^{a_j}}{\Gamma(a_j)} \int_0^\infty \cdots \int_0^\infty \frac{dx_j x_j^{a_j-1}}{[\mathcal{U}(x)]^{\frac{d}{2}}} \mathcal{U}(\partial) e^{i[\frac{-\mathcal{F}_0}{\mathcal{U}} - \sum_{i=1}^n x_i m_i^2]} =$$

$$i^l \left(\frac{\pi}{i}\right)^{d\frac{l}{2}} \prod_{j=1}^{n} \frac{i^{a_j}}{\Gamma(a_j)} \int_0^\infty \cdots \int_0^\infty \frac{dx_j x_j^{a_j-1}}{[\mathcal{U}(x)]^{\frac{d}{2}}} \mathcal{U}(x)(-i)^l e^{i[\frac{-\mathcal{F}_0}{\mathcal{U}} - \sum_{i=1}^n x_i m_i^2]}. \tag{2.81}$$

The resulting integral is proportional to the same integral with d changed to $d-2$

$$\mathcal{U}(\partial) I_G^d = (-\pi)^l I_G^{d-2}. \tag{2.82}$$

Now we can identify the masses with the occurring original masses.

After the tensor reduction, we get scalar integrals with various dimensions. However, we can write the IBP identities for an arbitrary but fixed dimension d. It means that we get master integrals with different dimensions. The task is to shift the dimension of all master integrals back to d. In order to do that we apply the method, which we introduced in this section, to master integrals. We assume that we have n master integrals $\{I_1, \cdots, I_n\}$. By applying eq.(2.82) to each master integral in $(d-2)$-dimension, and reducing the left-hand side to master integrals in d-dimensions, we obtain a system of equations

$$\mathcal{U}(\partial) I_i = \sum_{j=1}^{n} m_{ij} I_j, \tag{2.83}$$

where m_{ij} are coefficients containing the dimension d, the mass and the Mandelstam variables. In other words, we can express all these equations as follows

$$\begin{pmatrix} \mathcal{I}_1^{d-2} \\ \vdots \\ \mathcal{I}_n^{d-2} \end{pmatrix}_{master} = \underbrace{\mathbf{M}}_{n \times n \ Matrix} \times \begin{pmatrix} \mathcal{I}_1^{d} \\ \vdots \\ \mathcal{I}_n^{d} \end{pmatrix}_{master}. \tag{2.84}$$

In our case, however, all master integrals are in $(d+n)$-dimension ($n = 2, 4, 6, \cdots$) and we want to lower the dimension, i.e., we have to build the inverse matrix $\mathbf{M}^{-1}$. In practice it might be that the matrix $\mathbf{M}_{n \times n}$ can not be inverted. The reason for this is that the set of master integrals is not the minimal set, in other words, there are some integrals on the rhs. of eq.(2.84) which are not actually master integrals. This case could occur if we apply only IBP identities without considering the symmetries, especially in the sub-topologies. Sometimes we can use these relations in addition to IBP relations to find master integrals [37, 43, 47, 54].

There is another approach for the calculation of the shift in the dimension. This approach is based on analytic properties of Feynman integrals as functions of the parameter of dimensional regularization, d [5]. With this method we directly obtain the shift matrix which brings the set of master integrals in $(d+2)$-dimension into the set of the same master integrals in d.

We consider the basic parametric representation of a dimensionally regularized Feynman integral with h loops

$$I(a_1,\cdots,a_N,d) = \int \cdots \int \frac{d^d k_1 \cdots d^d k_h}{E_1^{a_1} \cdots E_N^{a_N}} \tag{2.85}$$

with $N = h(h+1)/2$ denominators, so that one can express every scalar product of loop momenta by them

$$E_r = \sum_{i \geq j} \mathcal{A}_r^{ij} k_i \cdot k_j - m_r^2 \tag{2.86}$$

and

$$k_l \cdot k_j = \sum_{r=1}^{N} (A_r^{-1})_{lj}(E_r + m_r^2). \tag{2.87}$$

We restrict ourselves at first to the vacuum case. However, we can generalize it to the non-vacuum case [55–59]. The IBP relations in the vacuum case originate from the following N equations

$$\int \cdots \int d^d k_1 \cdots d^d k_h \ \frac{\partial}{\partial k_i} \cdot \left(\frac{k_j}{E_1^{a_1} \cdots E_N^{a_N}} \right) = 0, \quad i \geq j. \tag{2.88}$$

By using eq.(2.86) and eq.(2.87) and the chain rule for

$$\frac{\partial}{\partial k_i} \cdot \left(\frac{k_j}{E_1^{a_1} \cdots E_N^{a_N}} \right),$$

eq.(2.88) can be represented in the following form

$$d\, \delta_{ij} I(a_1,\cdots,a_N,d) = 2 \sum_{a,r} \sum_{l}^{N\ h} \mathcal{A}_r^{il} a_r \mathcal{O}_r^+ (A_a^{-1})^{jl} (\mathcal{O}^- + m_r^2) I(a_1,\cdots,a_N,d), \tag{2.89}$$

where $\mathcal{O}_i^{\pm}I(\cdots, a_i, \cdots, d) = I(\cdots, a_i \pm 1, \cdots, d)$. Applying the relations

$$[a_i\mathcal{O}_i^+, \mathcal{O}_j^-] = \delta_{ij}, \qquad \sum_{r=1}^{N} \mathcal{A}_r^{il}(\mathcal{A}_r^{-1})^{jk} = \delta_{ij}\delta_{lk}, \tag{2.90}$$

leads to

$$\frac{d-h-1}{2}d\,\delta_{ij}I(a_1, \cdots, a_N, d) = \sum_{a,r}\sum_{l}^{N\ h}(\mathcal{A}_a^{-1})^{jl}(\mathcal{O}_a^- + m_a^2)\mathcal{A}_r^{il}a_r\mathcal{O}_r^+I(a_1, \cdots, a_N, d). \tag{2.91}$$

The final goal is to represent any scalar integral $I(a_1, \cdots, a_N, d)$ as a linear combination of master integrals $MI_i(d)$ with some coefficient functions $C^i(a_1, \cdots, a_N, d)$

$$I(a_1, \cdots, a_N, d) = \sum_i C^i(a_1, \cdots, a_N, d)MI_i(d). \tag{2.92}$$

The problem, to find a solution for eq.(2.91), is equivalent to constructing the coefficient functions C^i in eq.(2.92). We put eq.(2.92) in eq.(2.91) and so we get an equation in C^i. Note that it is a reformulation of the problem because the operators act on $C^i(a_1, \cdots, a_N, d)$ and not on $MI_i(d)$. If we find any set of the solutions, we could construct C^i as their linear combinations. We apply the following basic representation [57]

$$C^i(a_1, \cdots, a_N, d) = \frac{1}{(2\pi i)^N}\int\frac{dx_1\cdots dx_N}{x_1^{a_1}\cdots x_N^{a_N}}P(x_1 + m_1^2, \cdots, x_N + m_N^2)^{(d-h-1)/2}, \tag{2.93}$$

where the *Baikov polynomial* is

$$P(x_1, \cdots, x_N) = det\left(\sum_{a=1}^{N}(\mathcal{A}_a^{-1})^{jl}x_a\right). \tag{2.94}$$

Integration over the parameter x_i should be interpreted in a way that the IBP in this parametric integral is valid. In this case, such objects satisfy the initial IBP relations eq.(2.88)[60]. This property can be verified concerning the point that the operator $a_r\mathcal{O}^+$ can be replaced by the differential operator and the operator $\mathcal{O}^-$ by multiplication with x_r. We get the general solution for eq.(2.91)

$$C^i(a_1, \cdots, a_N, d) = \frac{1}{(2\pi i)^N}\int\frac{dx_1\cdots dx_N}{x_1^{a_1}\cdots x_N^{a_N}}det\left((\mathcal{A}_a^{-1})^{jl}(x_r + m_r^2)\right)^{(d-h-1)/2}. \tag{2.95}$$

If $C^i(a_1, \cdots, a_N, d)$ is a solution for eq.(2.91) then one can check that $P(\mathcal{O}^- + m^2)\,C^i(a_1, \cdots, a_N, d-2)$ is a solution, too [39, 57]. *Lee* [5] used the approach above to derive a relation between the loop integrals with different dimensions. To derive the *Lee* dimensional recurrence relation, we use the definition of a scalar integral as in eq. (2.85)

$$I(a_1, \cdots, a_N, d) = \int d^dk_1\cdots d^dk_h f(k_1, \cdots, k_h, p_1, \cdots, p_e), \tag{2.96}$$

where

$$f(k_1, \cdots, k_h, p_1, \cdots, p_e) = \frac{1}{E_1^{a_1} \cdots E_N^{a_N}}$$

and h is the number of loops and e the number of independent external momenta. The idea is now to split the integration measure into two independent parts. We consider the following simple integral

$$I(d) = \int \frac{d^d k}{k^2 (k - p)^2}, \quad p^2 \neq 0 \tag{2.97}$$

to illustrate the idea. The loop momentum k will be divided into the orthogonal and the parallel components $k_\perp$ and $k_{||}$ with the following conditions

$$\begin{aligned} k_{i\perp} \cdot q_j &= 0, \quad \forall q_j \in \{k_{i+1}, \cdots, k_h, p_1, \cdots, p_e\}, \\ k_{i\perp} \cdot k_{||} &= 0, \\ k &= k_{i\perp} + k_{||} \end{aligned} \tag{2.98}$$

$k_{||}$ can be written as a linear combination of the other momenta, in this case just p, i.e.,

$$k_{||} = a \cdot p \Rightarrow k_\perp = k - k_{||} = k - a \cdot p. \tag{2.99}$$

Using the first condition of eq. (2.98) we get

$$k_\perp \cdot p = 0 \Rightarrow (k - a \cdot p) \cdot p = k \cdot p - ap^2 = 0 \Rightarrow a = \frac{k \cdot p}{p^2}. \tag{2.100}$$

The property that $k_\perp$ occurs in every propagator in quadratic form, for instance

$$(k - p)^2 = p^2 + k_{||}^2 + k_\perp^2 - 2k_{||} \cdot p,$$

implies that we use the spherical representation of $k_\perp$.

We now present this idea in a more general form. The integration measure in eq.(2.96) is

$$d^d k_1 d^d k_2 \cdots d^d k_h = d^{M-1} k_{1||} d^{d-M+1} k_{1\perp} d^{M-2} k_{2||} d^{d-M+2} k_{2\perp} \cdots, \tag{2.101}$$

where every $k_{i||}$ lies in the subspace which is spanned by $k_{i+1}, \cdots, k_h, p_1, \cdots, p_e$. We have to recalculate the integration measure, i.e.,

$$d^{M-i} k_{i||} = \frac{ds_{i,i+1} ds_{i,i+2} \cdots ds_{i,M}}{\sqrt{G(k_{i+1}, \ldots, k_h, p_1, \ldots, p_e)}}, \quad M = h + e, \tag{2.102}$$

where s_{ij}[4] are the scalar products and

$$G(q_1, \cdots, q_n) = det(q_i \cdot q_j) \tag{2.103}$$

is the Gram determinant. For instance, we consider the following diagram:

---

[4]Here we consider the Euclidian case.

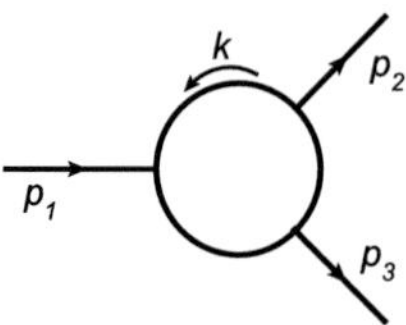

Figure 2.5.: The shift of the dimension with the help of the *Lee*-method [5]

Because of momentum conservation there are just two independent external momenta, p_1 and p_2, and now we can build the matrix

$$
\begin{array}{c|ccc}
 & k & p_1 & p_2 \\
\hline
k & k \cdot k & k \cdot p_1 & k \cdot p_2 \\
p_1 & p_1 \cdot k & p_1 \cdot p_1 & p_1 \cdot p_2 \\
p_2 & p_2 \cdot k & p_2 \cdot p_1 & p_2 \cdot p_2
\end{array}
\tag{2.104}
$$

The other measure $d^k k_{i\perp}$ can be written as

$$
d^n k_{i\perp} = \frac{1}{2} \Omega_n k_{i\perp}^{n-2} dk_{i\perp}^2 \, ,
\tag{2.105}
$$

when we integrate over the angle, where

$$
\Omega_n = \frac{2\pi^{n/2}}{\Gamma(n/2)}
\tag{2.106}
$$

is the n-dimensional full solid angle. $dk_{i\perp}^2$ can be replaced by ds_{ii}. The integration measure $dk_{i\perp}$ should be recalculated, too, i.e.,

$$
d^{d-M+i} k_{i\perp} = \frac{1}{2} \Omega_{d-M+i} \left(\frac{G(k_i, \cdots, k_h, p_1, \ldots, p_e)}{G(k_{i+1}, \cdots, k_h, p_1, \ldots, p_e)} \right)^{(d-M+i-2)/2} ds_{ii} \, ,
\tag{2.107}
$$

where h is the number of loops and e the number of independent external momenta and $i \in \{1, \cdots, h\}$. All the Gram determinants except the first and the last ones are canceled in the measure (2.101), and we get the following d–dimensional representation of a scalar integral [5]

$$
\int \frac{d^d k_1 \ldots d^d k_h}{\pi^{hd/2}} f = \frac{\pi^{-he/2 - h(h-1)/4}}{\prod_{i=1}^{h} \Gamma\left(\frac{d-M+i}{2}\right)}
$$

$$
\times \int \left(\prod_{i=1}^{h} \prod_{j=i}^{h+e} ds_{ij} \right) \frac{[G\left(k_1, \ldots k_h, p_1, \ldots, p_e\right)]^{(d-e-h-1)/2}}{[G\left(p_1, \ldots, p_e\right)]^{(d-e-1)/2}} f \, ,
\tag{2.108}
$$

where f is defined as in eq.(2.96). Now we can write eq. (2.108) as follows

$$I(a_1, \cdots, a_N, d) = \int \left(\prod_{i=1}^{h} \prod_{j=i}^{h+e} ds_{ij} \right) g(d, s) f, \tag{2.109}$$

where

$$g(d, s) = \frac{\pi^{-he/2 - h(h-1)/4}}{\prod_{i=1}^{h} \Gamma\left(\frac{d-M+i}{2}\right)} \frac{[G(k_1, \ldots k_h, p_1, \ldots, p_e)]^{(d-e-h-1)/2}}{[G(p_1, \ldots, p_e)]^{(d-e-1)/2}} \tag{2.110}$$

and s denotes the dependency of scalar products.

An integral in $(d+2)$-dimension is given by

$$
\begin{aligned}
I(a_1, \cdots, a_N, d+2) &= \int \left(\prod_{i=1}^{h} \prod_{j=i}^{h+e} ds_{ij} \right) g(d+2, s) f \\
&= \int \left(\prod_{i=1}^{h} \prod_{j=i}^{h+e} ds_{ij} \right) \frac{g(d+2, s)}{g(d, s)} \times g(d, s) f.
\end{aligned}
\tag{2.111}
$$

Our aim is to take $\frac{g(d+2,s)}{g(d,s)}$ out of the integral, however, that is not allowed because of the s-dependency of $\frac{g(d+2,s)}{g(d,s)}$. One can calculate $\frac{g(d+2,s)}{g(d,s)}$ by using eq.(2.108)

$$\frac{g(d+2, s)}{g(d, s)} = \frac{(2)^h \, [G(p_1, \ldots, p_e)]^{-1}}{\prod_{i=1}^{h}(d-e-h+i)} \left(G(k_1, \ldots k_h, p_1, \ldots, p_e) \right). \tag{2.112}$$

The Gram determinant $G(k_1, \ldots k_h, p_1, \ldots, p_e)$ contains scalar products which can be expressed by the propagators. That leads to the idea to replace this determinant by a polynomial of shift operators $P(\mathcal{O}^-)$

$$
\begin{aligned}
I(a_1, \cdots, a_N, d+2) &= \int \left(\prod_{i=1}^{h} \prod_{j=i}^{h+e} ds_{ij} \right) g(d+2, s) f \\
&= \int \left(\prod_{i=1}^{h} \prod_{j=i}^{h+e} ds_{ij} \right) \frac{(2)^h \, [G(p_1, \ldots, p_e)]^{-1}}{\prod_{i=1}^{h}(d-e-h+i)} P(\mathcal{O}^-) \times g(d, s) f \\
&= \frac{(2)^h \, [G(p_1, \ldots, p_e)]^{-1}}{\prod_{i=1}^{h}(d-e-h+i)} P(\mathcal{O}^-) \times \int \left(\prod_{i=1}^{h} \prod_{j=i}^{h+e} ds_{ij} \right) g(d, s) f \\
&\overset{5}{=} \frac{(2)^h \, [G(p_1, \ldots, p_e)]^{-1}}{\prod_{i=1}^{h}(d-e-h+i)} P(\mathcal{O}^-) I(a_1, \cdots, a_N, d).
\end{aligned}
\tag{2.113}
$$

---

[5] Using eq.(2.109).

So we obtain the lowering dimensional recurrence relation

$$I(a_1, \cdots, a_N, d+2) = \frac{(2)^h \left[G\left(p_1, \ldots, p_e \right) \right]^{-1}}{(d-e-h+1)_h} P(\mathcal{O}^-) I(a_1, \cdots, a_N, d), \tag{2.114}$$

where I is defined as in eq.(2.96). This is a practical formula for the computer implementation. However, if we use this formula to shift back the dimension of integrals we have to calculate some non-reducible scalar products, i.e. we need not just dots or just scalar products in the reduction tables but both of them. That might be difficult when we have to deal with double boxes.

2.3.3. Families of topologies

In multi-loop calculations we may generate many diagrams. Each diagram corresponds to a Feynman integral[6]. For the application of the IBP method to all these diagrams we need just the propagators and the Mandelstam variables. We can apply the IBP method to every diagram or alternatively we can find some topologies (diagrams) generating all other integrals by a shift of a loop momentum. Consider the case that we have master integrals and there is a list of already evaluated master integrals. Then it is extremely helpful if we can find our master integrals in the existing list, so we do not have to calculate these integrals. We can use the Feynman Graph Polynomials described in subsec. 2.3.1 to compare graphs with each other. The structure of every Feynman integral is determined by $\mathcal{U}$ and $\mathcal{F}$ which do not depend of the exponents of the denominator factors. That means we can use $\mathcal{U}$ and $\mathcal{F}$ to compare two diagrams. The only freedom that is left when we compare $\{\mathcal{U}_1, \mathcal{F}_1\}$ with $\{\mathcal{U}_2, \mathcal{F}_2\}$ is a permutation of lines, or equivalently, Schwinger parameters x_i. We have implemented an algorithm with Mathematica[61] which can compare two or more Feynman diagrams and possibly find the map between them. In practice it is better to use the products of $\mathcal{U}$ and $\mathcal{F}$ instead of the pairs $\{\mathcal{U}, \mathcal{F}\}$. Such an algorithm can be as follows:

- Build the $\mathcal{U}$ and $\mathcal{F}$ polynomials for each diagram using equations (2.75) and (2.76).

- Multiply $\mathcal{U}_i$ with $\mathcal{F}_i$ for $i \in \{1, 2\}$ (to compare diagram 1 with diagram 2).

- Make all possible permutations for Schwinger parameters in the product $\mathcal{U}_i \times \mathcal{F}_i$. In this step one gets a set in which there are all possible products of $\mathcal{U}_i$ with $\mathcal{F}_i$. In order to be able to compare the first diagram with the second one, in other words, to compare $\{\mathcal{U}_1 \times \mathcal{F}_1\}$ with $\{\mathcal{U}_2 \times \mathcal{F}_2\}$, we need the next step.

- Sort both sets $\{\mathcal{U}_1 \times \mathcal{F}_1\}_{(permutate)}$ and $\{\mathcal{U}_2 \times \mathcal{F}_2\}_{(permutate)}$. One can sort an algebraic set in several ways [62], for instance one can use the lexicographical order to sort a set. In this case, for the implementation with Mathematica [61], the Mathematica-Function *Sort*[$\cdots$] is used.

- Pick up the first element of each set for the comparison, if they are equal both diagrams belong to the same topology.

---

[6]We do not strictly distinguish between a diagram and the belonging Feynman integral.

It might happen, that two diagrams differ in the kinematical variables, s and t. In such cases we can repeat the algorithm after exchanging these in one of the diagrams. We implemented this algorithm in a Mathematica package [7].

---

[7]One can find this package and a minimal example at:
   `http://people.physik.hu-berlin.de/~assadsol`

3

QCD corrections to single top quark production at NNLO: One-loop squared contributions

As we can see in fig. 1.2 the NNLO calculation has many building blocks. In this chapter we begin with the calculation of the one-loop corrections squared. After a brief review of the importance of the single top quark production from the phenomenological point of view, we discuss in this chapter the status of the theory and the motivation of the NNLO calculation. For more detailed overviews of top quark physics, see [63–65]. Furthermore, we discuss some details of the calculation of the one-loop corrections squared to the single top quark production at NNLO.

3.1. Single Top Quark

In 1995, two experiments, CDF [66] and D$\emptyset$ [67, 68], at the Tevatron showed top-antitop quark pairs created from proton-antiproton collisions. The top quark is the heaviest known elementary particle. Because of its large mass, the top quark decays before hadronisation and thus gives a direct access to its properties such as spin polarization and charge.

In hadron collisions, like the proton-antiproton collisions at the Tevatron or the proton-proton collisions at the LHC, top quarks can be produced singly or in pairs. Top quark pair production occurs predominantly via the strong interaction, while single top quark production proceeds via the electroweak interaction. Single top quark production was first discovered at the Tevatron in 2009 [69, 70], much later than the top quark pair production because of the smaller cross sections as well as the larger backgrounds of single top quark production compared to top quark pair production. ATLAS and CMS, however, due to much larger cross sections available at the LHC, observed single top in 2011 [71, 72].

There are three main hadronic production modes in the SM for single top quark production, namely top quark production via the exchange of a virtual W boson in the t-channel and in the s-channel, and the associated production of a t quark and a real W boson, as one can see in fig. 3.1:

$$q\,b \to q'\,t\,, \quad q\,\bar{q}' \to \bar{b}\,t\,, \quad b\,g \to W^-\,t\,. \tag{3.1}$$

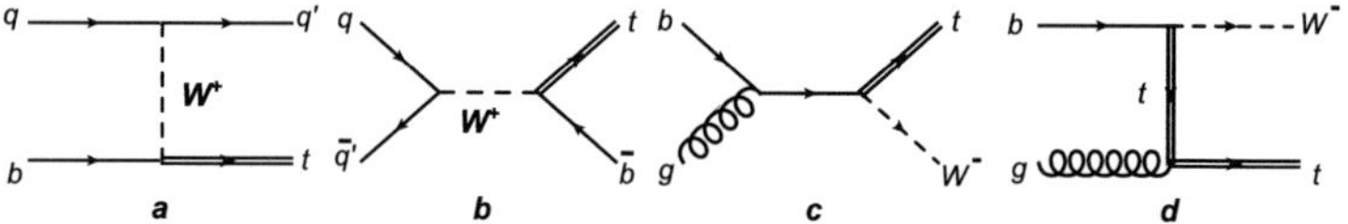

Figure 3.1.: The leading order Feynman diagrams for single top quark production processes: t-channel (a), s-channel (b), and associated tW production (c,d).

In each mode there is a charged-current interaction of the top quark in different kinematical regions. Since each mode has relatively distinct event kinematics they might be
observed separately. The experimental signatures of the s-channel and the t-channel are
different from each other. The t-channel production mode is the most important one for
the Tevatron and for the LHC because it has the largest cross section of the three channels
for both colliders. That makes up about 76% of the cross section at the LHC (7 TeV). At the
Tevatron (1.96 TeV) it is similar ($\sim$ 63%) [73, 74].

Studying single top quark production is interesting because of several reasons. First of
all, this enables a direct measurement of the Cabibbo-Kobayashi-Maskawa (CKM) matrix
element $|V_{tb}|$. It is possible to determine $|V_{tb}|$ indirectly from unitarity

$$|V_{ub}|^2 + |V_{cb}|^2 + |V_{tb}|^2 = 1$$

to a very high precision. This indirect determination assumes three generations of quarks
and the unitarity of the CKM matrix. The direct measurement, on the other hand, does
not depend of these assumptions. Secondly, the measurement of the spin polarization of
single top quarks can be used to test the $V - A$ structure of the weak charged-current
interaction. The top quark in single top quark production is produced left-handed. Since
top quarks decay before they can hadronize, spin correlations survive in the final decay
products [15, 16, 75–77]. Thirdly, from the experimental point of view the estimation of the
single top backgrounds is crucial to all signals with W+jet or $W + b$ as backgrounds. This
background is significant in a number of Higgs search channels [17] and new physics [19–
21]. Finally, the t-channel, shown in fig. 3.1, can be used to extract the b-quark density.

In the following subsection we review the theoretical results for the single top quark
production.

3.1.1. The theoretical status of the production cross section

The cross section for single top quark production in hadron collisions was calculated at
NLO in QCD in 2002 [14]. The most recent calculations also contain NNLL resummation [73, 78–81]. There are two studies constructing a NLO Monte Carlo calculation for
the t-channel [82] and the s-channel [83], focusing on signal cross sections and final state
kinematical distributions.

In tab. 3.1 and tab. 3.2 the LO and NLO cross sections are shown for the s-channel and
the t-channel single top quark production [14]. The difference between the two tables is

Process	$\sqrt{S}$	σ_{LO} (pb)	σ_{NLO} (pb)
s-channel	2 TeV $p\bar{p}$ (t)	0.315	0.463 $\pm$0.002
	14 TeV pp (t)	4.53	6.55 $\pm$0.03
t-channel	2 TeV $p\bar{p}$ (t)	0.948	1.029 $\pm$0.004
	14 TeV pp (t)	144.8	152.6 $\pm$0.6

Table 3.1.: σ_{LO} and σ_{NLO} for single top quark production at the Tevatron and LHC for $m_t = 175$ GeV. All scales set to m_t. Errors include only Monte Carlo statistics [14].

Process	$\sqrt{S}$	σ_{LO} (pb)	σ_{NLO} (pb)
s-channel	2 TeV $p\bar{p}$ (t)	0.297	0.459 $\pm$0.002
	14 TeV pp (t)	4.612	6.56 $\pm$0.03
t-channel	2 TeV $p\bar{p}$ (t)	1.068	1.062 $\pm$0.004
	14 TeV pp (t)	152.7	155.9 $\pm$0.6

Table 3.2.: σ_{LO} and σ_{NLO} for single top quark production at the Tevatron and LHC for $m_t = 175$ GeV. For the t-channel the scales are set to DDIS scales. Errors include only Monte Carlo statistics [14].

the different choice of the scales. In tab. 3.1, all scales are set to m_t, whereas in tab. 3.2 for the t-channel cross section the double deep-inelastic-scattering (DDIS) scales were chosen, μ_l^2 for the light quark line and μ_h^2 for the heavy quark line [84 86].

The deviation between σ_{LO} and σ_{NLO} in tab. 3.2 is less than 3%. One might come to the conclusion that we do not need to consider any corrections beside NLO. At LO and NLO there is no exchange of a gluon between two quark lines because of the colour conservation. However, this is no longer the case at NNLO. By comparison of the values of σ_{LO} and σ_{NLO} in tab. 3.1 with the same values in tab. 3.2, we can recognize that the deviation between σ_{LO} and σ_{NLO} in tab. 3.1 is greater than in tab. 3.2. The reason for that is the various choice of the scales. For the calculation of the t-channel cross section one can choose double deep-inelastic-scattering (DDIS) scales, since the t-channel process can be factorized into two independent corrections, each of which resembles deep-inelastic-scattering [84]. But that can not be continued readily at NNLO. All these properties exist at NLO but not at NNLO [1]. Now the question arises as to whether one will have a similar accuracy for the calculation of the cross section when one takes into account the higher order corrections. In order to verify this rather good precision at NLO, we have to go to higher order corrections. A full NNLO calculation has many building blocks as one can see in fig. 1.2. In this work, we mainly concentrate on virtual contributions which are one-loop corrections squared and two-loop QCD corrections. In the following sections we will discuss the one-loop corrections squared.

---

[1] For a detailed discussion of the scale variation of the σ_{LO} and σ_{NLO} for single top quark production one can see [14].

3.2. General setup

There are six one-loop Feynman diagrams to single top quark production at NNLO in the
t-channel. These diagrams were generated by using QGRAF [87]. There are two vertex
diagrams (shown in fig. 3.2) and four box diagrams (shown in fig. 3.3).

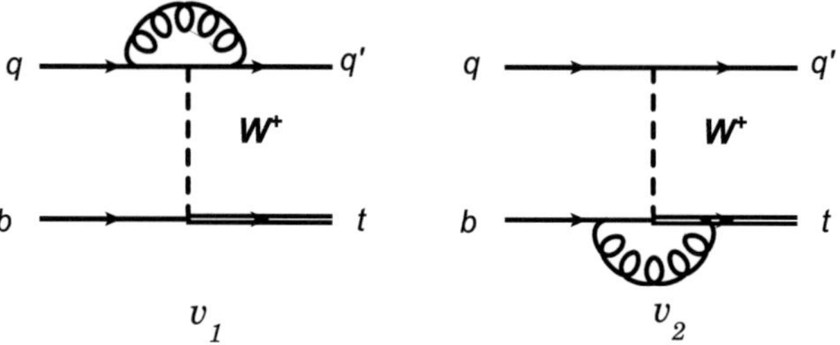

Figure 3.2.: Vertex corrections to the t-channel at NNLO

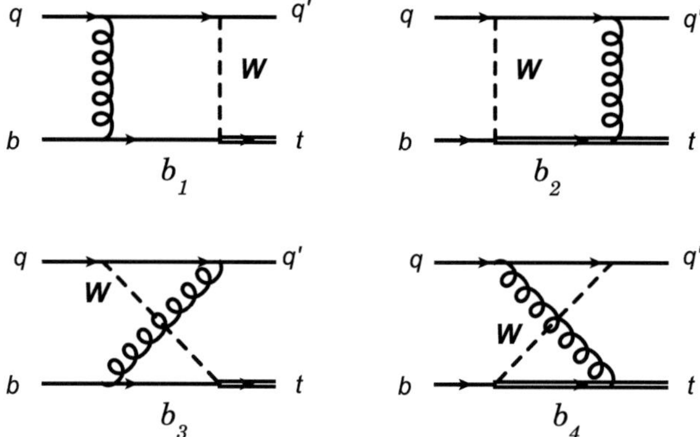

Figure 3.3.: Box corrections to the t-channel at NNLO

The one-loop corrections at NNLO contain two parts, the first one is the interference
between box diagrams and the second part is the interference between vertex corrections.
Because of the colour, the interference between the vertex diagrams and the box diagrams
does not contribute to these corrections.

In this calculation the propagator belonging to the W boson is used in the Feynman
gauge. To handle divergences occurring at intermediate stages of the calculation, we use
dimensional regularization.

For one-loop diagrams, one has to deal with the tensor integrals which occur. We used the Tarasov-method in subsec. 2.1.2 to reduce the tensor integrals to scalar integrals. We can use the other method, the projection method subsec. 2.1.3, for the reduction of tensor integrals too. Since we have no diagrams containing closed fermion triangle loops we can use the naive scheme for the treatment of γ_5 [88]. However, a naive approach keeping the anti-commuting property of γ_5 in d dimensions leads to ambiguities. The problem is that the cyclicity of the Dirac trace like

$$\varepsilon^{\mu\nu\rho\sigma} Tr(\gamma_\tau \gamma_\mu \gamma_\nu \gamma_\rho \gamma_\sigma \gamma^\tau) \tag{3.2}$$

is no longer valid [89–91]. We can alternatively use the scheme described by *Larin* in [92] for the treatment of the axial current

$$J_\mu^{5a}(x) = \overline{\psi}(x)\, \gamma_\mu \gamma_5 t^a\, \psi(x), \tag{3.3}$$

where $\psi(x)$ is a quark field and t^a is a generator of a flavor group.

Using the naive scheme for the treatment of γ_5, we get the following spin structures for the t-channel[2]

$$
\begin{aligned}
S_1 &= \overline{u}(q_1)\, \gamma_7\, u(p_2)\overline{u}(q_2)\, \gamma_6\slashed{q}_1\, u(p_1) \\
S_2 &= \overline{u}(q_1)\, \gamma_6\slashed{p}_1\, u(p_2)\overline{u}(q_2)\, \gamma_6\slashed{q}_1\, u(p_1) \\
S_3 &= \overline{u}(q_1)\, \gamma_6\gamma_{\mu_1}\, u(p_2)\overline{u}(q_2)\, \gamma_6\gamma_{\mu_1}\, u(p_1) \\
S_4 &= \overline{u}(q_1)\, \gamma_7\gamma_{\mu_1}\slashed{p}_1\, u(p_2)\overline{u}(q_2)\, \gamma_6\gamma_{\mu_1}\, u(p_1) \\
S_5 &= \overline{u}(q_1)\, \gamma_7\gamma_{\mu_1}\gamma_{\mu_2}\, u(p_2)\overline{u}(q_2)\, \gamma_6\gamma_{\mu_1}\gamma_{\mu_2}\slashed{q}_1\, u(p_1) \\
S_6 &= \overline{u}(q_1)\, \gamma_6\gamma_{\mu_1}\gamma_{\mu_2}\slashed{p}_1\, u(p_2)\overline{u}(q_2)\, \gamma_6\gamma_{\mu_1}\gamma_{\mu_2}\slashed{q}_1\, u(p_1) \\
S_7 &= \overline{u}(q_1)\, \gamma_6\gamma_{\mu_1}\gamma_{\mu_2}\gamma_{\mu_3}\, u(p_2)\overline{u}(q_2)\, \gamma_6\gamma_{\mu_1}\gamma_{\mu_2}\gamma_{\mu_3}\, u(p_1),
\end{aligned} \tag{3.4}
$$

where γ_{μ_i} ($i = 1, \cdots, 4$) are the gamma matrices, $\gamma_6 = 1 + \gamma_5$ and $\gamma_7 = 1 - \gamma_5$. $u(p_i)$ and $\overline{u}(q_i)$ ($i = 1, 2$) are the Dirac spinors, where p_1 and p_2 are the momenta of the incoming quarks, q and b, respectively, and q_1 and q_2 the momenta of the outgoing quarks, t and q'. In the vertex diagrams only S_1 and S_3 occur, whereas in the box diagrams all spin structures arise.

We used both schemes, naive and *Larin*, for the treatment of γ_5 to calculate the contribution of the one-loop squared.

3.3. Results

After the tensor reduction, we get integrals which might be UV-divergent. By the application of the Passarino-Veltman reduction [26], the C_{24}-coefficient in this calculation is the only one with a UV-divergency [93]. The amplitudes belonging to the vertex diagrams can be written as follows

---

[2]We get similar spin structures for the s-channel.

$$\mathcal{A}_{v_1} = c_0 \left((m_t^2 - s - t)[-2C_0(v_1, 1, 2, 3) - 2C_{11} + C_{22}(d-2) - C_{23}(d-2)] + C_{24}(d-2)^2 \right) S_3,$$

$$\mathcal{A}_{v_2} = c_0 \left(-4C_{12}m_t - 2C_{23}(d-2)m_t \right) S_1 + c_0 \left(2(s+t)C_0(v_2, 1, 2, 3) + 2m_t^2 C_{12} \right.$$

$$\left. + C_{22}(d-2)(m_t^2 - s - t) + 2C_{11}(s+t) + C_{23}(d-2)(s+t) + C_{24}(d-2)^2 \right) S_3, \tag{3.5}$$

where v_1 and v_2 are the labels shown in fig. 3.2 and the C_{ij}, $i, j \in \{1, 2, 3, 4\}$ are the Passarino-Veltman coefficients as in [26]. c_0 is given by

$$c_0 = \frac{\alpha_e \alpha_s V_{tb} V_{ud}^*}{8 \sin^2 \theta_W (t - m_w^2)}, \tag{3.6}$$

where α_e is the fine-structure constant and α_s is the coupling constant of the strong interaction. θ_W indicates the Weinberg angle. V_{tb} and V_{ud}^* are the elements of the CKM-matrix and m_w is the mass of the W boson.

In the naive scheme we can perform the calculation in the following way, we do the renormalization in $\overline{MS}$-scheme before squaring the amplitude

$$\mathcal{A}_{\mathcal{R}} = \mathcal{A} - Z_{\overline{MS}} \times Born, \tag{3.7}$$

where $\mathcal{A}$ is the amplitude, $\mathcal{A}_{\mathcal{R}}$ is the renormalized amplitude and $Z_{\overline{MS}}$ is the counter term of the $\overline{MS}$-scheme. The renormalized amplitudes in $d = 4$ for the vertex diagrams can be written as

$$\mathcal{A}_{\mathcal{R},v_1} = c_0 \left(-2 - 3\overline{B}_0(v_1, 2, 3) + 2(s + t - m_t^2)C_0(v_1, 1, 2, 3) \right) S_3 \, ,$$

$$\mathcal{A}_{\mathcal{R},v_2} = c_0 \left(\frac{2m_t}{s+t} + \frac{2}{m_t(s+t)} \overline{A}(v_2, 3) - \frac{4m_t}{s+t} \overline{B}_0(v_2, 1, 3) + \frac{2m_t}{s+t} \overline{B}_0(v_2, 2, 3) \right) S_1$$

$$+ c_0 \left(-2 + \frac{2(-m_t^2 + s + t)}{s+t} \overline{B}_0(v_2, 1, 3) + \frac{2m_t^2 - 3(s+t)}{s+t} \overline{B}_0(v_2, 2, 3) + 2(s+t)C_0(v_2, 1, 2, 3) \right) S_3, \tag{3.8}$$

where t and s are the Mandelstam variables

$$p_1 \cdot p_2 = \frac{s}{2}, \qquad p_2 \cdot q_2 = -\frac{t}{2}, \tag{3.9}$$

m_t is the top mass, $\mathcal{A}_{v_1}$ belongs to the diagram in fig. 3.2(a) and $\mathcal{A}_{v_2}$ belongs to the diagram in fig. 3.2(b). In eq. (3.8)

$$\overline{B}_0(v_i, 1, 3) = B_0(v_i, 1, 3) - \frac{1}{\epsilon},$$

$$\overline{B}_0(v_i, 2, 3) = B_0(v_i, 2, 3) - \frac{1}{\epsilon}, \qquad i \in 1, 2$$

$$\overline{A}(v_2, 3) = A(v_2, 3) - \frac{m_t^2}{\epsilon}, \tag{3.10}$$

denote UV-subtracted master integrals, where $A(v_2, 3)$, $B_0(v_i, 1, 3)$, $B_0(v_i, 2, 3)$ and $C_0(v_i, 1, 2, 3)$ for $i \in \{1, 2\}$ are the master integrals as defined in app. A.1. Then we build the product of $\mathcal{A}_\mathcal{R} \mathcal{A}_\mathcal{R}^\dagger$ in four dimensions.

Since the box diagrams in fig. 3.3 have no UV-divergency, one can treat them with the naive scheme for γ_5 and then square the sum of boxes. Since the final results of the box diagrams are rather large in terms of master integrals we present them in app. A.2. The final results were calculated in two independent calculations and with two different methods, described in subsec. 2.1.1 and in subsec. 2.1.2. The final results reached with the two methods and the two calculations are consistent with each other. In eq.(3.2) we mentioned the problem with the treatment of γ_5 in the naive scheme. This motivated us to do this calculation in another scheme. We have chosen the scheme described by *Larin* [92] for the treatment of the axial current

$$\gamma_\mu \gamma_5 \to Z_5^{ns} \frac{i}{3!} \, \varepsilon_{\mu \mu_1 \mu_2 \mu_3} \gamma^{\mu_1} \gamma^{\mu_2} \gamma^{\mu_3}, \tag{3.11}$$

where

$$Z_5^{ns} = 1 - \frac{\alpha_s}{\pi} C_F + \mathcal{O}(\alpha_s^2), \tag{3.12}$$

where C_F is the Casimir operator of the defining representation of the colour group. This compact notation is favourable for an implementation of this scheme with a computer algebra system. For the calculation in this scheme we used the tensor reduction method described in subsec. 2.1.2. Using this scheme, we get more spin structures and the intermediate expressions are getting huge. The final result calculated in this scheme is too long to be presented here. As a consistency check, we set $d = 4$ and $Z_5^{ns} = 1$ in the Larin scheme, and found a full agreement with the result calculated in the naive scheme at $d = 4$. This comparison is not trivial because we calculated the one-loop squared contributions in each scheme with different setups. In the naive scheme we used the Clifford-Algebra, the Dirac equation and the projectors, eq.(3.4), whereas we did not use all these in the Larin scheme.

In general, one compares the UV-divergent parts of the results in the naive and the *Larin* schemes with each other. Since the IR-divergences in the real corrections contribute to the same pole structures, a complete comparison between both schemes can only be done when we consider the real contributions, too. We compared the coefficients of the UV-divergent parts with each other and found a complete agreement.

4

QCD form factors of heavy quarks at NNLO: Axial vector contributions

In chapter 3 we calculated the one-loop squared corrections to the single top quark production at NNLO. In order to check our setup and verify independently the results in the literature [3], we calculate the axial vector form factors. After a brief introduction to QCD form factor we calculate the axial vector form factors excluding the anomalous triangle contributions. The final results in terms of master integrals are rather large and so we can not present them here. However, we compared the results for the vector coupling with [94, 95] and for the axial vector coupling with [3]. Our results agree with these results.

4.1. QCD form factor

The vertex function of an amplitude $\mathcal{A}^{\mu}(V \to q\bar{q})$ ($V = \gamma^{*}, Z$) (see fig. 4.1) can be decomposed into so-called form factors, where q and $\bar{q}$ are the on-shell massive quark and antiquark with the momenta p_1 belonging to q and p_2 belonging to $\bar{q}$ and the mass m

$$\mathcal{A}^{\mu}(V \to q\bar{q}) = \bar{u}(p_1) \; \Gamma^{\mu}(p_3) \; v(p_2) \,, \tag{4.1}$$

where $p_3 = p_1 + p_2$.

$$\Gamma^{\mu}(p_3) = (-i)\Big(g_V F_1(s)\gamma^{\mu} \;\; + g_V \frac{1}{2m} F_2(s) i\sigma^{\mu\nu} p_{3_\nu}$$
$$+ g_A G_1(s)\gamma^{\mu}\gamma_5 + g_A \frac{1}{2m} G_2(s)\gamma_5 p_3^{\mu}\Big), \tag{4.2}$$

where $s = (p_1 + p_2)^2/m^2$, $\sigma^{\mu\nu} = \frac{i}{2}[\gamma^{\mu}, \gamma^{\nu}]$ and g_V, g_A are the vector and the axial vector couplings.

The study of the form factor is both of phenomenological interest, for example the forward-backward asymmetry in the production of bottom quarks [96, 97], and of theoretical interest. The electromagnetic form factor of heavy quarks as a gauge invariant building block represents the simplest example of a scattering amplitude in QCD. The perturbative amplitude for the process $V \to q\bar{q}$, $\mathcal{A}(V \to q\bar{q})$ ($V = \gamma^{*}, Z$) expanded in the strong coupling constant α_s is represented by:

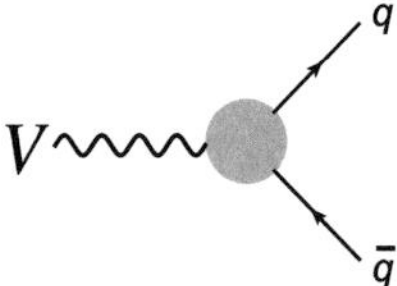

Figure 4.1.: Form factor: The gray circle represents the sum over all possible higher order QCD corrections

There are explicit results available for the QCD corrections at NNLO for the vector coupling [94, 95] as well as for the axial-vector coupling [3] and the anomaly contributions [98].

4.2. The axial vector form factors

$\mathcal{A}^\mu(p_2, q_1)$ is the vertex amplitude for the decay of a virtual Z boson, with incoming momentum $p_1 = q_1 - p_2$, into a massive quark-antiquark pair of momenta p_2 und q_1:

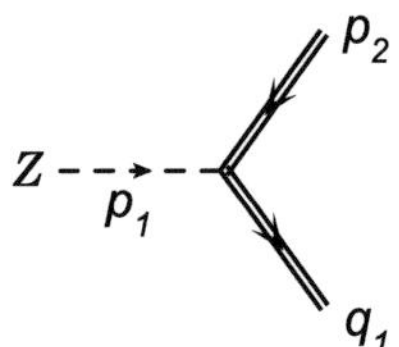

Figure 4.2.: Axial vector form factors

The Z boson has the following coupling

$$\frac{ig}{2\cos\theta_w}\gamma_\mu(g_V - g_A\gamma_5). \tag{4.3}$$

In the case of $g_V = 1$ and $g_A = 0$ we have the vector coupling, whereas for $g_V = 0$ and $g_A = 1$ we have the axial vector coupling.

We generated in total 12 diagrams with QGRAF [87], represented in fig. 4.3, excluding the anomalous triangle diagrams, shown in fig. 4.4, contributing to the two-loop QCD corrections. There are three mirrored diagrams to *axd2, axd4* and *axd6* [1].

---

[1] The reason for the numbering of the diagrams in fig. 4.3 is that in our calculation the diagram *axd1* is the most complicated one and has the highest number of master integrals. That is the reason why this diagram is the first diagram in the numbering. Then in turn, each diagram is numbered up to mirrored diagrams to *axd2, axd4* and *axd6*. The mirrored diagrams are numbered by *asd3, asd5* and *asd7*.

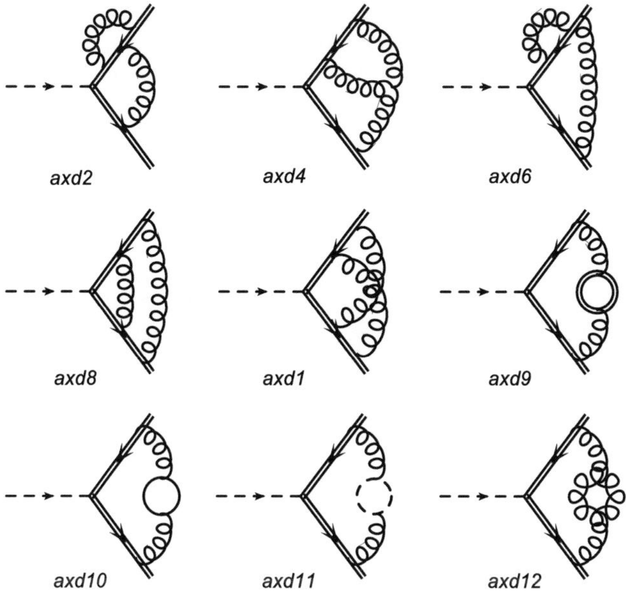

Figure 4.3.: Two-loop QCD axial vector: The straight lines refer to massless quarks and the dashed lines to ghosts. The double straight line represents the massive quark.

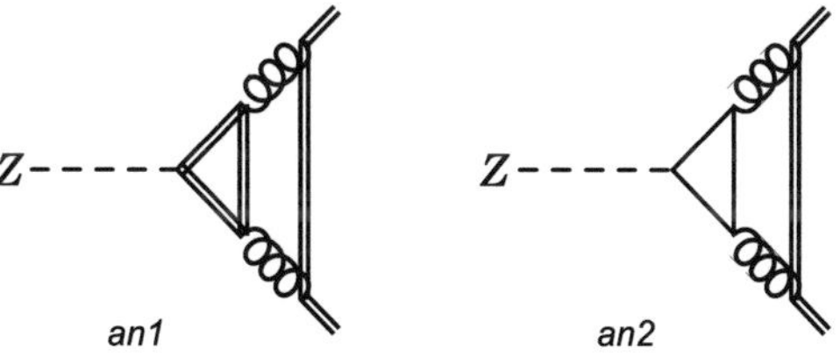

Figure 4.4.: Triangle diagrams: The double straight lines represent the massive quark. The straight lines refer to massless quarks.

In this calculation we use an anticommuting γ_5 in d dimensions. Since we do not consider the contribution of the diagrams containing closed triangle loops, we will not have any anomaly in the final result [88]. When one uses another prescription for γ_5 there will be no difference at the end of the calculation in the final results [92, 99]. For this calculation we used both, the projection method described in subsec. 2.1.3, and the Tarasov-method explained in subsec. 2.1.2. To apply the projection method we need the spin structures, which we use as projectors. After we applied the naive scheme for the treatment of γ_5 and Dirac algebra, we got the following general spin structures

$$
\begin{aligned}
S_1 &= \bar{u}(q_1)\ \gamma_6\ v(p_2)\ p_2^\mu, & S_2 &= \bar{u}(q_1)\ \gamma_7\ v(p_2)\ p_2^\mu, \\
S_3 &= \bar{u}(q_1)\ \gamma_6\ v(p_2)\ q_1^\mu, & S_4 &= \bar{u}(q_1)\ \gamma_7\ v(p_2)\ q_1^\mu, \\
S_5 &= \bar{u}(q_1)\ \gamma_6\ \gamma_\mu\ v(p_2), & S_6 &= \bar{u}(q_1)\ \gamma_7\ \gamma_\mu\ v(p_2),
\end{aligned}
\tag{4.4}
$$

where γ_μ is the gamma matrix, $\gamma_6 = 1 + \gamma_5$ and $\gamma_7 = 1 - \gamma_5$ as in eq.(3.4). $u(q_1)$ and $v(p_2)$ are the Dirac spinors with the momenta p_2 (incoming) and q_1 (outgoing). We calculated the tensor coefficients for the application of the projection method, subsec. 2.1.3, in app. D.1.

The spin structures in eq.(4.4) contain the left and right chiral projection operators, $1 - \gamma_5$ and $1 + \gamma_5$. The following transformations help us to divide the spin structures into parts containing γ_5 and parts not containing it

$$
\begin{aligned}
S_1 + S_2 &+ (S_3 + S_4) = S_1' \\
S_1 + S_2 &- (S_3 + S_4) = S_2' \\
S_1 - S_2 &+ (S_3 - S_4) = S_3' \\
S_1 - S_2 &- (S_3 - S_4) = S_4' \\
S_5 + S_6 &= S_p, \quad S_5 - S_6 = S_m.
\end{aligned}
\tag{4.5}
$$

We solve these equations for $S_1, \cdots, S_6$ and obtain the following replacement rules

$$
\begin{aligned}
S_1 &= \frac{1}{4}\left(S_1' + S_2' + S_3' + S_4'\right), & S_2 &= \frac{1}{4}\left(S_1' + S_2' - S_3' - S_4'\right), \\
S_3 &= \frac{1}{4}\left(S_1' - S_2' + S_3' - S_4'\right), & S_4 &= \frac{1}{4}\left(S_1' - S_2' - S_3' + S_4'\right), \\
S_5 &= \frac{1}{2}\left(S_p + S_m\right), & S_6 &= \frac{1}{2}\left(S_p - S_m\right).
\end{aligned}
\tag{4.6}
$$

Since two spin structures are invariant under a CP transformation we find just S_1', S_4' , S_m and S_p in the sum of all diagrams occurring in fig. 4.3. The number of Feynman diagrams in this calculation is relatively small so that we do not need to use the representative topologies which are introduced in subsec. 2.3.3. For the reduction to master integrals we used Reduze2 [25]. We applied the same setup as in chapter 3. Thus, we restrict ourselves to master integrals. Almost all master integrals occurring in the sum of all diagrams in fig. 4.3 can be obtained from the diagram $axd1$ in fig. 4.3. The corresponding scalar integral to this diagram is

$$
\begin{aligned}
I_{axd1}(d, a_1, a_2, a_3, a_4, a_5, a_6) = \int\int d^d k_1 d^d k_2 \frac{1}{(k_1^2 - m_t)^{a_1}(k_2^2 - m^2)^{a_2}((p_1 - k_1)^2 - m^2)^{a_3}} \\
\times \frac{1}{((-p_2 + k_2)^2)^{a_4}((k_1 + k_2)^2)^{a_5}((q_1 - k_1 - k_2)^2 - m^2)^{a_6}}.
\end{aligned}
\tag{4.7}
$$

We complete the set of propagators with the last propagator $(k_1 - p_2)^2$ to get all possible scalar products, i.e., we consider the following scalar integral

$$I_{axd1}(d, a_1, a_2, a_3, a_4, a_5, a_6, a_7) = \int \int d^d k_1 d^d k_2 \frac{1}{(k_1^2 - m_t)^{a_1}(k_2^2 - m^2)^{a_2}((p_1 - k_1)^2 - m^2)^{a_3}}$$

$$\times \frac{1}{((-p_2 + k_2)^2)^{a_4}((k_1 + k_2)^2)^{a_5}((q_1 - k_1 - k_2)^2 - m^2)^{a_6}((k_1 - p_2)^2)^{a_7}}. \tag{4.8}$$

The master integrals belonging to this diagram are

$$
\begin{aligned}
MI_{axd1,1} &= I_{axd1}(d, 1, 1, 0, 0, 0, 0, 0) & MI_{axd1,2} &= I_{axd1}(d, 1, 1, 1, 0, 0, 0, 0) \\
MI_{axd1,3} &= I_{axd1}(d, 1, 1, 0, 0, 0, 1, 0) & MI_{axd1,4} &= I_{axd1}(d, 1, 0, 0, 1, 1, 0, 0) \\
MI_{axd1,5} &= I_{axd1}(d, 0, 1, 1, 0, 1, 0, 0) & MI_{axd1,6} &= I_{axd1}(d, -1, 1, 1, 0, 1, 0, 0) \\
MI_{axd1,7} &= I_{axd1}(d, 1, 1, 1, 0, 0, 1, 0) & MI_{axd1,8} &= I_{axd1}(d, 1, 1, 1, -1, 0, 1, 0) \\
MI_{axd1,9} &= I_{axd1}(d, 1, 1, 0, 1, 0, 1, 0) & MI_{axd1,10} &= I_{axd1}(d, 1, 1, -1, 1, 0, 1, 0) \\
MI_{axd1,11} &= I_{axd1}(d, 1, 0, 1, 1, 1, 0, 0) & MI_{axd1,12} &= I_{axd1}(d, 0, 1, 1, 1, 1, 0, 0) \\
MI_{axd1,13} &= I_{axd1}(d, 1, 1, 1, 1, 1, 0, 0) & MI_{axd1,14} &= I_{axd1}(d, 1, 1, 1, 1, 1, -1, 0) \\
MI_{axd1,15} &= I_{axd1}(d, 1, 1, 1, 1, 1, 1, 0) & MI_{axd1,16} &= I_{axd1}(d, 1, 1, 1, 1, 1, 1, -1) \cdot
\end{aligned}
\tag{4.9}
$$

We can find these integrals expressed in terms of harmonic polylogarithms (HPLs) in the literature [95]. For this aim we have to change the basis of the integral space in order to get the master integrals occurring in [95]. In this representation, the master integrals with scalar products, negative powers $(a_i, i = 1, \cdots, 7)$, are expressed by master integrals with dots in eq. (2.28). There are some possibilities to do this change of basis, we can for instance write down the corresponding IBP relations and then solve these for the integrals which we want to choose as a basis. But for the reduction to master integrals we use Reduze2 [25]. This software helps us to do this, too. We have to write an export file with integrals which we want to have as master integrals in app. D.2. We chose the following integrals as master integrals

$$
\begin{aligned}
GI_{axd1,1} &= I_{axd1}(d, 0, 0, 1, 0, 0, 1, 0) & GI_{axd1,2} &= I_{axd1}(d, 1, 0, 1, 0, 0, 1, 0) \\
GI_{axd1,3} &= I_{axd1}(d, 0, 1, 1, 0, 0, 1, 0) & GI_{axd1,4} &= I_{axd1}(d, 0, 1, 1, 0, 1, 0, 0) \\
GI_{axd1,5} &= I_{axd1}(d, 0, 1, 2, 0, 1, 0, 0) & GI_{axd1,6} &= I_{axd1}(d, 0, 0, 1, 1, 1, 0, 0) \\
GI_{axd1,7} &= I_{axd1}(d, 1, 1, 1, 0, 0, 1, 0) & GI_{axd1,8} &= I_{axd1}(d, 1, 1, 1, 0, 0, 2, 0) \\
GI_{axd1,9} &= I_{axd1}(d, 0, 1, 1, 0, 1, 1, 0) & GI_{axd1,10} &= I_{axd1}(d, 0, 1, 1, 0, 1, 2, 0) \\
GI_{axd1,11} &= I_{axd1}(d, 1, 0, 1, 1, 1, 0, 0) & GI_{axd1,12} &= I_{axd1}(d, 0, 1, 1, 1, 1, 0, 0) \\
GI_{axd1,13} &= I_{axd1}(d, 1, 0, 1, 1, 1, 1, 0) & GI_{axd1,14} &= I_{axd1}(d, 2, 0, 1, 1, 1, 1, 0) \\
GI_{axd1,15} &= I_{axd1}(d, 1, 1, 1, 1, 1, 1, 0) & GI_{axd1,16} &= I_{axd1}(d, 1, 1, 2, 1, 1, 1, 0).
\end{aligned}
\tag{4.10}
$$

Now, one can use the expansion of these integrals in terms of HPLs in the literature [95].

4.2.1. Summary

Since the final result in terms of master integrals is rather large we do not state it here [2]. We calculated the QCD form factors of heavy quarks with two different methods described in subsec. 2.1.2 and subsec. 2.1.3. The final results reached by the two methods are consistent with each other. We found a complete agreement between our results and the results in the literature for the vector coupling [94, 95] as well as for the axial vector coupling [3]. Our calculation for $G_1(s)$ and $G_2(s)$ in eq.(4.2) is the first independent check of the calculation which is done in [3].

This consistency check has ensured us even more that our setup for such a calculation is correct, before we begin with the calculation of the two-loop contributions to single top quark production. Because for the next step, the calculation of the double box contributions to single top production, it is rather complicated to check the result with the help of two methods and two independent calculations.

---

[2]One can download the final result depending on the HPLs from:
http://people.physik.hu-berlin.de/~assadsol

5

Two-loop QCD corrections to single top quark production

In the previous chapters we used two methods described in subsec. 2.1.2 and subsec. 2.1.3 to reduce the tensor integrals. Using each method for the reduction of the tensor integrals leads to another starting point for the application of the IBP method. We used both methods in chapter 3 to calculate the contribution of the one-loop corrections squared to single top quark production at NNLO. We concentrate in this chapter on the two-loop contributions which are divided into two parts: First, vertex contributions which we discuss in section 5.2. We apply the methods described in chapter 2 to the two-loop vertex diagrams of the single top quark production. For the reduction of the tensor integrals we can use both methods described in subsec. 2.1.2 and subsec. 2.1.3. Secondly, double boxes which we study in section 5.3. Compared to vertex diagrams there is one more propagator in the double box diagrams and we have one more variable in the problem, the mass of the W boson. Therefore, the calculation of the double boxes is more complicated compared to the calculation of the vertex diagrams. The methods which we use for this calculation are the same as described in chapter 2. However, the application of the tensor reduction explained in subsec. 2.1.2 makes the next step, i.e., solving the IBP identities, more difficult (see section 5.3). In subsec. 5.3.1 we discuss some subtle technical details which are applied to make the reduction to master integrals possible. The number of variables in the problem, s, t, m_t, m_w and d, makes not only the reduction to master integrals more difficult but the whole calculation. We can replace the mass of the W boson m_w by a rather exact numerical ratio between the mass of the top quark and the mass of the W boson. So we can calculate all diagrams with this numerical approximation.

5.1. General setup

In the calculation of the two-loop corrections several diagrams occur. The contribution of the two-loop corrections to the amplitude at NNLO is the sum of all such diagrams

$$Amplitude_{2loops} = \sum_{i=1}^{n} \mathcal{D}_i, \tag{5.1}$$

where D_i represents the i-th diagram. Each diagram can be decomposed in a colour structure c_i and a Lorentz structure L_i. We can do the calculation separately for both structures because they factorize

$$\sum_{i=1}^{n} \mathcal{D}_i = \sum_{i=1}^{n} c_i L_i. \tag{5.2}$$

We can use the Fierz identity

$$(t^a)^i_k (t^a)^l_j = \frac{1}{2} \delta^i_j \delta^l_k - \frac{1}{2N_c} \delta^i_k \delta^l_j \tag{5.3}$$

to calculate the colour coefficients, where t^a are the generators of the fundamental representation of $SU(N)$. We multiply the two-loop diagrams with the colour structure of the Born amplitude: $\delta_{lj}\,\delta_{ik}$. The contribution of some two-loop diagrams vanishes because of the vanishing colour factors. We get the following colour factors when we multiply the two-loop diagrams with the Born amplitude: $I_2(R)C_F N^2$, $I_2(R)C_F N$, $C_A C_F N^2$ and $C_F^2 N^2$, where C_F and C_A are the Casimir-Operators and N denotes the dimension of the fundamental representation of $SU(N)$. $I_2(R)$ sets the normalization of the representation R. These colour factors divide the diagrams into gauge invariant sub-groups. We generated 121 two-loop diagrams with QGRAF[87] for the t- and s-channel. After multiplication with the Born amplitude and calculation of the colour factors, just 47 diagrams remain. There are 29 vertex correction diagrams and 18 double boxes. In the next step we attempt to identify all occurring topologies. For this goal we use the algorithm which is described in the previous chapter, in subsec. 2.3.3. We apply the IBP method to these topologies in order to find the master integrals. In total, there are 14 topologies. That means we have to reduce just 14 diagrams and then we can map the other diagrams onto these 14 diagrams by a shift of the loop momenta in subsec. 2.3.3. Each topology is represented by a diagram. Five topologies, enumerated as $v1, v2, v3, v4, v5$, belong to the vertex corrections, shown in fig. 5.1, and nine topologies to the double boxes, shown in fig. 5.3 and fig. 5.2[1].

We use the representing integrals of these 14 topologies for the application of the IBP method. There are 4 sub-topologies of double boxes belonging to the vertex topologies, see fig. 5.4. We sort the topologies by colour factors to get gauge invariant partial sums of eq.(5.2). One can see the colour factors in tab. 5.1.
The sum of all vertex topologies $v1, v2, v3, v4, v5$, and $Sd1, Sd2, Sd7, Sd8$ can be considered as a gauge invariant part as well as the sum of all double boxes $d1, \cdots, d9$.
Now we have to deal with tensor structures occurring in every diagram. We presented two methods in the previous chapter: first, the Tarasov-method in subsec. 2.1.2, and second, the projection method in subsec. 2.1.3. For the vertex corrections we used both methods. To handle divergences occurring at intermediate stages of the calculation we use dimensional regularization. We use the naive scheme for the treatment of γ_5 because we do not have any anomaly [88]. Before we begin with the tensor reduction, we determine

---

[1]See section 5.3 for the double boxes.

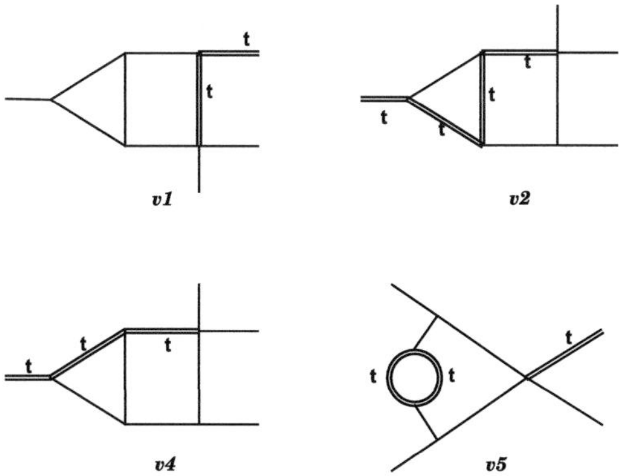

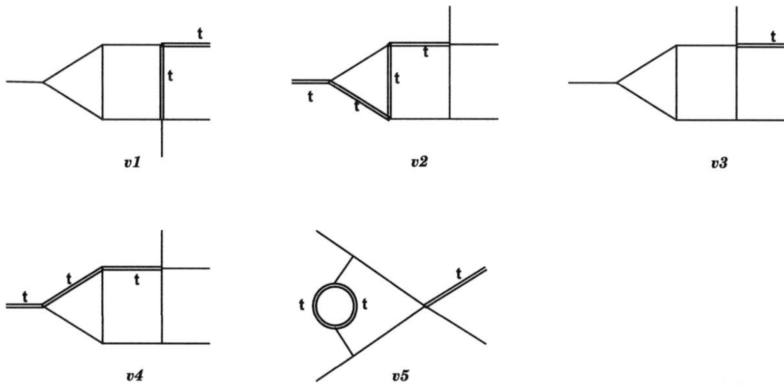

Figure 5.1.: Vertex corrections topologies. The straight lines refer to massless quarks and the double straight line represents the top quark.

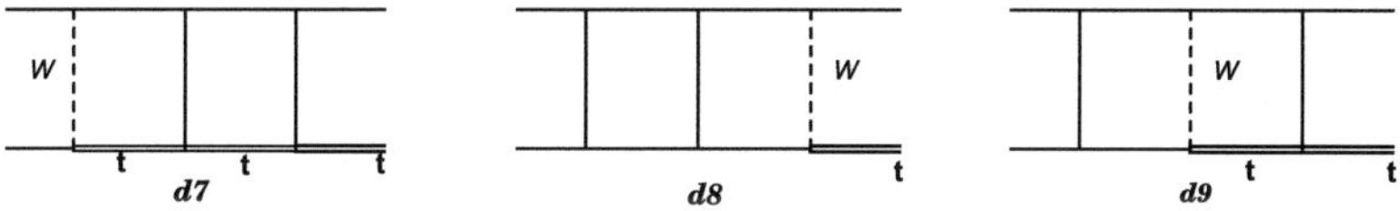

Figure 5.2.: Planar double box topologies. The straight lines refer to massless quarks and the double straight line represents the top quark.

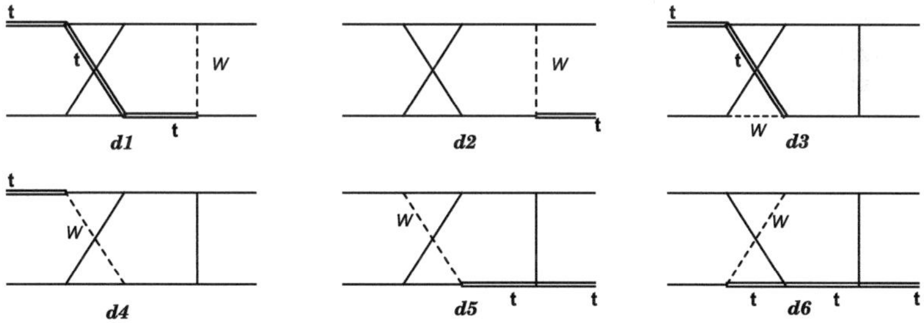

Figure 5.3.: Non-planar double box topologies. The straight lines refer to massless quarks and the double straight line represents the top quark.

the spin structures. We get the general spin structures for the two-loop QCD corrections

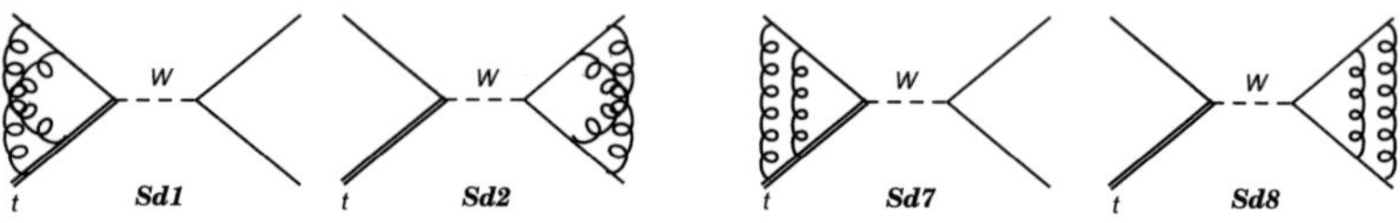

Figure 5.4.: Sub-topologies of double boxes. The straight lines refer to massless quarks and the double straight line represents the top quark.

Nr.	Colour factor	Topology
1	$I_2(R)C_F N^2$	$v3, v4, v5$
2	$I_2(R)C_F N$	$d1, d2, d3, d4, d5, d6, d7, d8, d9$
3	$C_A C_F N^2$	$Sd1, Sd2, v1, v2, v3, v4$
4	$C_F^2 N^2$	$Sd1, Sd2, Sd7, Sd8, v1, v2, v3, v4$

Table 5.1.: Colour factors belonging to the topologies

$(t$-channel$)^2$ by using the Dirac algebra and the momentum conservation

$$
\begin{aligned}
S_1 &= \bar{u}(q_1)\,\gamma_7\,u(p_2)\bar{u}(q_2)\,\gamma_6\slashed{q}_1\,u(p_1) \\
S_2 &= \bar{u}(q_1)\,\gamma_6\slashed{p}_1\,u(p_2)\bar{u}(q_2)\,\gamma_6\slashed{q}_1\,u(p_1) \\
S_3 &= \bar{u}(q_1)\,\gamma_6\gamma_{\mu_1}\,u(p_2)\bar{u}(q_2)\,\gamma_6\gamma_{\mu_1}\,u(p_1) \\
S_4 &= \bar{u}(q_1)\,\gamma_7\gamma_{\mu_1}\slashed{p}_1\,u(p_2)\bar{u}(q_2)\,\gamma_6\gamma_{\mu_1}\,u(p_1) \\
S_5 &= \bar{u}(q_1)\,\gamma_7\gamma_{\mu_1}\gamma_{\mu_2}\,u(p_2)\bar{u}(q_2)\,\gamma_6\gamma_{\mu_1}\gamma_{\mu_2}\slashed{q}_1\,u(p_1) \\
S_6 &= \bar{u}(q_1)\,\gamma_6\gamma_{\mu_1}\gamma_{\mu_2}\slashed{p}_1\,u(p_2)\bar{u}(q_2)\,\gamma_6\gamma_{\mu_1}\gamma_{\mu_2}\slashed{q}_1\,u(p_1) \\
S_7 &= \bar{u}(q_1)\,\gamma_6\gamma_{\mu_1}\gamma_{\mu_2}\gamma_{\mu_3}\,u(p_2)\bar{u}(q_2)\,\gamma_6\gamma_{\mu_1}\gamma_{\mu_2}\gamma_{\mu_3}\,u(p_1) \\
S_8 &= \bar{u}(q_1)\,\gamma_7\gamma_{\mu_1}\gamma_{\mu_2}\gamma_{\mu_3}\slashed{p}_1\,u(p_2)\bar{u}(q_2)\,\gamma_6\gamma_{\mu_1}\gamma_{\mu_2}\gamma_{\mu_3}\,u(p_1) \\
S_9 &= \bar{u}(q_1)\,\gamma_7\gamma_{\mu_1}\gamma_{\mu_2}\gamma_{\mu_3}\gamma_{\mu_4}\,u(p_2)\bar{u}(q_2)\,\gamma_6\gamma_{\mu_1}\gamma_{\mu_2}\gamma_{\mu_3}\gamma_{\mu_4}\slashed{q}_1\,u(p_1) \\
S_{10} &= \bar{u}(q_1)\,\gamma_6\gamma_{\mu_1}\gamma_{\mu_2}\gamma_{\mu_3}\gamma_{\mu_4}\slashed{p}_1\,u(p_2)\bar{u}(q_2)\,\gamma_6\gamma_{\mu_1}\gamma_{\mu_2}\gamma_{\mu_3}\gamma_{\mu_4}\slashed{q}_1\,u(p_1) \\
S_{11} &= \bar{u}(q_1)\,\gamma_6\gamma_{\mu_1}\gamma_{\mu_2}\gamma_{\mu_3}\gamma_{\mu_4}\gamma_{\mu_5}\,u(p_2)\bar{u}(q_2)\,\gamma_6\gamma_{\mu_1}\gamma_{\mu_2}\gamma_{\mu_3}\gamma_{\mu_4}\gamma_{\mu_5}\,u(p_1),
\end{aligned}
\tag{5.4}
$$

where γ_{μ_i} $(i = 1, \cdots, 4)$ are the gamma matrices and γ_6, γ_7, $u(p_i)$ and $\bar{u}(q_i)$ $(i = 1, 2)$ as well as the external momenta are defind as in eq. (3.4). We keep the full spin dependency because the top quark in the single top quark production is polarized. S_1 and S_3 occur in the vertex corrections.

5.2. Vertex contributions

In this section we apply the tensor reduction methods described in subsec. 2.1.2 and subsec. 2.1.3. After the application of the first method, the dimension of master integrals should be shifted back. For this purpose we used two methods, the Tarasov-method and

---

[2]We obtain similar spin structures for the s-channel.

the Lee-method, which we explained in chapter 2. Finally, we use the projection method described in subsec. 2.1.3 to calculate the vertex contributions.

5.2.1. Calculation of vertex diagrams

At first we applied the Tarasov-method for the tensor reduction. We implemented the algorithm described in subsec. 2.1.2. Since we have two loops and tensors of the fourth rank we get up to 8 *dots*. The increasing of the power of each propagator by one is called a *dot*, see eq. (2.28). After tensor reduction, the integrals will be shifted in the dimension. The higher the number of *dots*, the more difficult it is to apply the IBP method. We use Reduze1 [48] and Reduze2 [25] to get master integrals. After getting all master integrals we apply the method explained in subsec. 2.3.2 to shift the dimension of the integrals back to d.

To illustrate this method with the help of a concrete calculation, we take the simplest occurring topology, namely topology $v3$. The scalar integral to this topology is

$$I_{v3}(d, a_1, a_2, a_3, a_4, a_5, a_6) = \int \frac{d^d k_1}{(2\pi)^d} \frac{d^d k_2}{(2\pi)^d} \frac{1}{((-k_2)^2)^{a_1} ((k_1)^2)^{a_2} ((-p_1 + k_1)^2)^{a_3}}$$
$$\times \frac{1}{((-p_2 + k_2)^2)^{a_4} ((-p_1 + k_1 - k_2)^2)^{a_5} ((-q_1 - q_2 + k_1)^2)^{a_6}}, \tag{5.5}$$

where k_1 and k_2 are the loop momenta and the momenta p_i (incoming) and q_i (outgoing) are the external momenta. q_1 is the momentum belonging to the top quark.

In the first step we determine the master integrals of topology $v3$. This toplogy has two master integrals $MI_{v3, 1}(d)$ and $MI_{v3, 2}(d)$

$$MI_{v3, 1}(d) = I_{v3}(d, 0, 2, 0, 1, 1, 0),$$
$$MI_{v3, 2}(d) = I_{v3}(d, 2, 1, 0, 0, 1, 1), \tag{5.6}$$

where the numbers in parentheses denote the powers of the propagators. Then we have to write down the first Symanzik polynomial $\mathcal{U}$ (see eq. (2.76)). Schwinger parameters act as increasing operators, i.e., a Schwinger parameter x_i increases the power of the i-th propagator like in eq.(2.82)

$$\mathcal{U} \times MI_{v3, 1}(d, 0, 2, 0, 1, 1, 0) = I_{v3}(d + 2, 0, 2, 0, 2, 2, 0) + 2I_{v3}(d + 2, 0, 3, 0, 1, 2, 0)$$
$$+ 2I_{v3}(d + 2, 0, 3, 0, 2, 1, 0),$$
$$\mathcal{U} \times MI_{v3, 2}(d, 2, 1, 0, 0, 1, 1) = I_{v3}(d + 2, 2, 1, 0, 0, 2, 2) + I_{v3}(d + 2, 2, 2, 0, 0, 2, 1)$$
$$+ 2I_{v3}(d + 2, 3, 1, 0, 0, 1, 2) + 2I_{v3}(d + 2, 3, 1, 0, 0, 2, 1)$$
$$+ 2I_{v3}(d + 2, 3, 2, 0, 0, 1, 1). \tag{5.7}$$

In the next step we have to reduce all scalar integrals on the right hand side to master integrals and then invert the matrix as explained in subsec. 2.3.2, because we need a shift formula from $d + 2$ to d. At the end we obtain the following formulas for the shift of the dimension

$$MI_{v3,\,1}(d+2) = \frac{(d-4)s^2}{3(d-3)(3d-8)(3d-4)}\,MI_{v3,\,1}(d),$$

$$MI_{v3,\,2}(d+2) = \frac{-4(d-4)s^2}{3(d-2)(3d-10)(3d-8)}MI_{v3,\,2}(d), \tag{5.8}$$

where $s = (p_1 + p_2)^2$ is the Mandelstam variable. We can calculate the formulas for the shift of the dimension for the other topologies in the same way. For the next topologies we will just give the end results. The scalar integral belonging to topology $v1$ is given by

$$I_{v1}(d,a_1,a_2,a_3,a_4,a_5,a_6) = \int \frac{d^dk_1}{(2\pi)^d}\,\frac{d^dk_2}{(2\pi)^d}\,\frac{1}{(-k_1^2 - m_t^2)^{a_1}\,(k_2^2)^{a_2}\,((q_1+k_1)^2)^{a_3}}$$

$$\times \frac{1}{((-p_1 - p_2 - k_1)^2)^{a_5}\,((q_2 + k_2)^2)^{a_4}\,((-q_1 - k_1 + k_2)^2)^{a_6}}. \tag{5.9}$$

There are four master integrals

$$MI_{v1,\,1}(d) = I_{v1}(d,1,1,0,0,0,1), \qquad MI_{v1,\,2}(d) = I_{v1}(d,1,-1,0,1,0,1),$$

$$MI_{v1,\,3}(d) = I_{v1}(d,1,0,0,1,0,1), \qquad MI_{v1,\,4}(d) = I_{v1}(d,1,1,0,0,1,1). \tag{5.10}$$

The final formulas for the shift of the dimension of topology $v1$ are

$$MI_{v1,\,1}(2+d) = \frac{-16\,(2\,d-5)\,(2\,d-3)\,m_t^4}{3\,(d-2)\,(d-1)\,(3\,d-4)\,(3\,d-2)}MI_{v1,\,1}(d),$$

$$MI_{v1,\,2}(2+d) = -\left(\frac{-((d-2)\,d\,m_t^6) + 12\,(d-2)\,(d-1)\,m_t^4\,s + 3\,(8 + d\,(17\,d-30))\,m_t^2\,s^2}{3\,(d-1)\,d\,(3\,d-4)\,(3\,d-2)\,s}\right.$$

$$\left. + \frac{2\,(d-2)\,d\,s^3}{3\,(d-1)\,d\,(3\,d-4)\,(3\,d-2)\,s}\right)MI_{v1,\,2}(d)$$

$$- \left(\frac{(m_t^2 - s)^2\,((d-2)^2\,m_t^4 + 2\,(4 + d\,(7\,d-13))\,m_t^2\,s + (d-2)\,d\,s^2)}{3\,(d-1)\,d\,(3\,d-4)\,(3\,d-2)\,s}\right)MI_{v1,\,3}(d),$$

$$MI_{v1,\,3}(2+d) = -\left(\frac{2\,((d-2)\,m_t^4 + 2\,(-10+7\,d)\,m_t^2\,s + (d-2)\,s^2)}{(-8 + 26\,d - 27\,d^2 + 9\,d^3)\,(m_t^2 - s)}\right)MI_{v1,\,2}(d)$$

$$- \left(\frac{(m_t^2 - s)^2\,((-10+7\,d)\,m_t^2 + (d-2)\,s)}{(-8 + 26\,d - 27\,d^2 + 9\,d^3)\,(m_t^2 - s)}\right)MI_{v1,\,3}(d),$$

$$MI_{v1,\,4}(2+d) = \left(\frac{(m_t^2 + s)((-2+d)m_t^4 + 2(7-3d)m_t^2 s + (-2+d)s^2)}{3\,(d-1)\,(3\,d-8)\,(3\,d-4)\,s^2}\right)MI_{v1,\,1}(d)$$

$$- \frac{4\,(d-3)\,(m_t^2 - s)^4}{3\,(d-1)\,(3\,d-8)\,(3\,d-4)\,s^2}\,MI_{v1,\,4}(d). \tag{5.11}$$

The scalar integral belonging to topology $v2$ is

$$I_{v2}(d, a_1, a_2, a_3, a_4, a_5, a_6) = \int \frac{d^d k_1}{(2\pi)^d} \frac{d^d k_2}{(2\pi)^d} \frac{1}{((-k_2)^2 - m_t^2)^{a_1} (k_1^2)^{a_2} ((q_1 + k_2)^2)^{a_3}}$$

$$\times \frac{1}{((q_2 + k_1)^2)^{a_4} ((p_1 + p_2 + k_1)^2 - m_t^2)^{a_5} ((q_2 + k_1 - k_2)^2 - m_t^2)^{a_6}}, \quad (5.12)$$

and there are 10 master integrals for this topology

$$\begin{aligned}
MI_{v2,1}(d) &= I_{v2}(d, -1, 1, 1, 0, 0, 1), & MI_{v2,6}(d) &= I_{v2}(d, 1, 1, -1, 0, 1, 1), \\
MI_{v2,2}(d) &= I_{v2}(d, 0, 0, 1, 1, 0, 1), & MI_{v2,7}(d) &= I_{v2}(d, 1, 1, 0, 0, 1, 0), \\
MI_{v2,3}(d) &= I_{v2}(d, 0, 1, 1, 0, 0, 1), & MI_{v2,8}(d) &= I_{v2}(d, 1, 1, 0, 0, 1, 1), \\
MI_{v2,4}(d) &= I_{v2}(d, 1, 0, 0, 0, 1, 0), & MI_{v2,9}(d) &= I_{v2}(d, 1, 1, 1, -1, 0, 1), \\
MI_{v2,5}(d) &= I_{v2}(d, 1, 0, 0, 0, 1, 1), & MI_{v2,10}(d) &= I_{v2}(d, 1, 1, 1, 0, 0, 1).
\end{aligned} \quad (5.13)$$

The final formulas for the dimension shift of topology $v2$ are

$$MI_{v2,1}(2+d) = \left(\frac{((d-2)d\,m_t^6 - 12(d-2)(d-1)m_t^4 s - 3(8+d(17d-30))m_t^2 s^2)}{3(d-1)d(3d-4)(3d-2)} \right.$$

$$\left. + \frac{(-2(d-2)d\,s^3)}{3(d-1)d(3d-4)(3d-2)} \right) MI_{v2,1}(d)$$

$$+ \frac{(m_t^4 - s^2)((d-2)^2 m_t^4 + 2(4+d(7d-13))m_t^2 s + (d-2)d\,s^2)}{3(d-1)d(3d-4)(3d-2)s} MI_{v2,3}(d),$$

$$MI_{v2,2}(2+d) = \frac{-16(2d-5)(2d-3)m_t^4}{3(d-2)(d-1)(3d-4)(3d-2)} MI_{v2,2}(d),$$

$$MI_{v2,3}(2+d) = \frac{2(-2m_t^4 + d\,m_t^4 - 20m_t^2 s + 14d\,m_t^2 s - 2s^2 + d\,s^2)}{(d-1)(3d-4)(3d-2)(m_t^2 + s)} MI_{v2,1}(d)$$

$$- \frac{(m_t^2 - s)(-10m_t^2 + 7d\,m_t^2 - 2s + d\,s)}{(d-1)(3d-4)(3d-2)} MI_{v2,3}(d),$$

$$MI_{v2,4}(2+d) = \frac{(4m_t^4)}{d^2} MI_{v2,4}(d),$$

$$MI_{v2,5}(2+d) = \frac{2(-16+11d)m_t^2}{(d-1)(3d-4)(3d-2)} MI_{v2,4}(d) + \frac{64(d-2)m_t^4}{3(d-1)(3d-4)(3d-2)} MI_{v2,5}(d),$$

$$MI_{v2,6}(2+d) = \left(\frac{2(d-3)(d-2)(d-1)d\,m_t^{12} + (d-3)(d-2)(24+17(d-2)d)m_t^{10} s}{3(d-3)(d-2)(d-1)d(3d-4)(3d-2)m_t^2 (m_t^2 - s)^2 s^2} \right.$$

$$+ \frac{d(14+d(-3+d(23d-54)))m_t^8 s^2 - (d-1)(662d - 311d^2 + 50d^3 - 432)m_t^6 s^3}{3(d-3)(d-2)(d-1)d(3d-4)(3d-2)m_t^2 (m_t^2 - s)^2 s^2}$$

$$+ \frac{(d-2)(-144+d(267+d(34d-161)))m_t^4 s^4 + (d-2)d(5+d(5d-12))m_t^2 s^5}{3(d-3)(d-2)(d-1)d(3d-4)(3d-2)m_t^2 (m_t^2 - s)^2 s^2}$$

$$\left. + \frac{(d-2)^2(d-1)d\,s^6}{3(d-3)(d-2)(d-1)d(3d-4)(3d-2)m_t^2 (m_t^2 - s)^2 s^2} \right) MI_{v2,4}(d)$$

$$+2\,m_t^2\,(m_t^2+s)\left(\frac{(d-3)(d-2)(d-1)\,m_t^8+4\,(d-3)(d-2)\,d\,m_t^6\,s}{3\,(d-2)^2\,(d-1)\,(3\,d-4)\,(3\,d-2)\,(m_t^2-s)^2\,s^2}\right.$$

$$+\frac{-2\,(14+d\,(27+d\,(13\,d-46)))\,m_t^4\,s^2+4\,(d-3)(d-2)\,d\,m_t^2\,s^3}{3\,(d-2)^2\,(d-1)\,(3\,d-4)\,(3\,d-2)\,(m_t^2-s)^2\,s^2}$$

$$\left.+\frac{(d-3)(d-2)(d-1)\,s^4}{3\,(d-2)^2\,(d-1)\,(3\,d-4)\,(3\,d-2)\,(m_t^2-s)^2\,s^2}\right)MI_{v2,\,5}(d)$$

$$-4\left(\frac{(d-2)^2\,(d-1)\,m_t^{12}+(d-2)\,d\,(3\,d-5)\,m_t^{10}\,s-(d-2)\,(30+d\,(9\,d-47))\,m_t^8\,s^2}{3\,(d-2)^2\,(d-1)\,(3\,d-4)\,(3\,d-2)\,(m_t^2-s)^2\,s^2}\right.$$

$$+\frac{2\,(-16+d\,(3\,d-5)(7\,d-18))\,m_t^6\,s^3-(d-2)\,(30+d\,(9\,d-47))\,m_t^4\,s^4}{3\,(d-2)^2\,(d-1)\,(3\,d-4)\,(3\,d-2)\,(m_t^2-s)^2\,s^2}$$

$$\left.+\frac{(d-2)\,d\,(3\,d-5)\,m_t^2\,s^5+(d-2)^2\,(d-1)\,s^6}{3\,(d-2)^2\,(d-1)\,(3\,d-4)\,(3\,d-2)\,(m_t^2-s)^2\,s^2}\right)MI_{v2,\,6}(d)$$

$$+8\,m_t^2\left(\frac{-2\,(d-3)(d-2)(d-1)\,m_t^8+(d-2)\,(24+5\,d\,(3\,d-5))\,m_t^6\,s}{3\,(d-2)^2\,d\,(8+9\,(d-2)\,d)\,(m_t^2-s)^2}\right.$$

$$+\frac{2\,(36+d\,(-72+(31-5\,d)\,d))\,m_t^4\,s^2+(d-2)\,(24+5\,d\,(3\,d-5))\,m_t^2\,s^3}{3\,(d-2)^2\,d\,(8+9\,(d-2)\,d)\,(m_t^2-s)^2}$$

$$\left.+\frac{-2\,(d-3)(d-2)(d-1)\,s^4}{3\,(d-2)^2\,d\,(8+9\,(d-2)\,d)\,(m_t^2-s)^2}\right)MI_{v2,\,7}(d)$$

$$+4\,m_t^2\,(m_t^2+s)^2\left(\frac{(d-2)(d-1)\,m_t^8+8\,(d-2)\,m_t^6\,s+2\,(-6+d\,(7\,d-17))\,m_t^4\,s^2}{3\,(d-2)^2\,(8+9\,(d-2)\,d)\,(m_t^2-s)^2\,s^2}\right.$$

$$\left.+\frac{8\,(d-2)\,m_t^2\,s^3+(d-2)(d-1)\,s^4}{3\,(d-2)^2\,(8+9\,(d-2)\,d)\,(m_t^2-s)^2\,s^2}\right)MI_{v2,\,8}(d),$$

$$MI_{v2,\,7}(2+d)=\frac{m_t^2\,(m_t^2+s)}{(d-1)\,d\,s}\,MI_{v2,\,4}(d)-\frac{m_t^2\,(m_t^2-s)^2}{(d-1)\,d\,s}\,MI_{v2,\,7}(d),$$

$$MI_{v2,\,8}(2+d)=\left(\frac{2\,(d-3)(d-2)\,m_t^8+(6+d\,(8\,d-21))\,m_t^6\,s+(d-4)(3\,d-4)\,m_t^4\,s^2}{3\,(d-3)(d-2)(d-1)(3\,d-4)\,s\,(m_t^3-m_t\,s)^2}\right.$$

$$\left.+\frac{(d-2)(2\,d-1)\,m_t^2\,s^3+(d-2)^2\,s^4}{3\,(d-3)(d-2)(d-1)(3\,d-4)\,s\,(m_t^3-m_t\,s)^2}\right)MI_{v2,\,4}(d)$$

$$+2\,m_t^2\,(m_t^2+s)\left(\frac{(d-3)(d-2)\,(m_t^4+s^2)-2\,(16+d\,(5\,d-19))\,m_t^2\,s}{3\,(d-2)^2\,(d-1)\,(3\,d-4)\,(m_t^2-s)^2\,s}\right)MI_{v2,\,5}(d)$$

$$-4\left(\frac{(d-2)^2\,(m_t^8+s^4)+4\,(d-2)\,(m_t^6\,s+m_t^2\,s^3)+2\,(24+7\,(d-4)\,d)\,m_t^4\,s^2}{3\,(d-2)^2\,(d-1)\,(3\,d-4)\,(m_t^2-s)^2\,s}\right)MI_{v2,\,6}(d)$$

$$+\left(\frac{4\,m_t^2\,(5\,(d-2)\,m_t^4+2\,(-10+3\,d)\,m_t^2\,s+5\,(d-2)\,s^2)}{3\,(d-2)^2\,(3\,d-4)\,(m_t^2-s)^2}\right)MI_{v2,\,7}(d)$$

$$+\left(\frac{4\,m_t^2\,(m_t^2+s)^2\,((d-2)\,m_t^4+2\,(3\,d-8)\,m_t^2\,s+(d-2)\,s^2)}{3\,(d-2)^2\,(3\,d-4)\,(m_t^2-s)^2\,s}\right)MI_{v2,\,8}(d),$$

$$
MI_{v2,\,9}(2+d) = -2\left(\frac{(d-2)\,d^2\,(3\,d-8)\,m_t^8 + d\,(-48+(d-2)\,d\,(42\,d-121))\,m_t^6\,s}{(d-2)\,(d-1)\,d\,(3\,d-4)\,(3\,d-2)\,((3\,d-8)\,m_t^2+(d-2)\,s)\,(m_t^4-s^2)}\right.
$$

$$
+\frac{(d-2)\,(-72+d\,(226+d\,(4\,d-129)))\,m_t^4\,s^2}{(d-2)\,(d-1)\,d\,(3\,d-4)\,(3\,d-2)\,((3\,d-8)\,m_t^2+(d-2)\,s)\,(m_t^4-s^2)}
$$

$$
\left.+\frac{(d-2)\,(-24+d\,(76+d\,(14\,d-59)))\,m_t^2\,s^3+(d-2)^2\,(d-1)\,d\,s^4}{(d-2)\,(d-1)\,d\,(3\,d-4)\,(3\,d-2)\,((3\,d-8)\,m_t^2+(d-2)\,s)\,(m_t^4-s^2)}\right)MI_{v2,\,1}(d)
$$

$$
+8\,m_t^4\left(\frac{d^2\,(1+d)\,m_t^4+2\,(d-2)\,(3+d\,(7\,d-8))\,m_t^2\,s}{(3\,(d-2)\,d\,(3\,d-4)\,(3\,d-2)\,((3\,d-8)\,m_t^2+(d-2)\,s)}\right.
$$

$$
\left.+\frac{(d-3)\,(d-2)\,d\,s^2}{(3\,(d-2)\,d\,(3\,d-4)\,(3\,d-2)\,((3\,d-8)\,m_t^2+(d-2)\,s)}\right)MI_{v2,\,10}(d)
$$

$$
+\left(\frac{d\,(-48+d\,(350+d\,(71\,d-323)))\,m_t^8}{3\,(d-2)\,(d-1)\,d\,(3\,d-4)\,(3\,d-2)\,(m_t^2-s)\,((3\,d-8)\,m_t^2+(d-2)\,s)}\right.
$$

$$
+\frac{2\,(360+d\,(-1258+d\,(1311+d\,(75\,d-533))))\,m_t^6\,s}{3\,(d-2)\,(d-1)\,d\,(3\,d-4)\,(3\,d-2)\,(m_t^2-s)\,((3\,d-8)\,m_t^2+(d-2)\,s)}
$$

$$
+\frac{6\,(d-2)\,(12+d\,(-17+10\,(d-1)\,d))\,m_t^4\,s^2}{3\,(d-2)\,(d-1)\,d\,(3\,d-4)\,(3\,d-2)\,(m_t^2-s)\,((3\,d-8)\,m_t^2+(d-2)\,s)}
$$

$$
+\frac{-2\,(d-2)\,(-24+d\,(69+d\,(11\,d-51)))\,m_t^2\,s^3}{3\,(d-2)\,(d-1)\,d\,(3\,d-4)\,(3\,d-2)\,(m_t^2-s)\,((3\,d-8)\,m_t^2+(d-2)\,s)}
$$

$$
\left.+\frac{-3\,(d-2)^2\,(d-1)\,d\,s^4}{3\,(d-2)\,(d-1)\,d\,(3\,d-4)\,(3\,d-2)\,(m_t^2-s)\,((3\,d-8)\,m_t^2+(d-2)\,s)}\right)MI_{v2,\,3}(d)
$$

$$
-4\,m_t^2\left(\frac{(432+d\,(-1458+d\,(1979+d\,(-1264+(371-40\,d)\,d))))\,m_t^6}{3\,(d-3)\,(d-2)\,(d-1)\,d\,(3\,d-4)\,(3\,d-2)\,(m_t^2-s)^2\,((3\,d-8)\,m_t^2+(d-2)\,s)}\right.
$$

$$
+\frac{(d-2)\,(12+d\,(-263+d\,(474+d\,(56\,d-291))))\,m_t^2\,s^2+(d-3)^2\,(d-2)\,(3\,d-4)\,s^3}{3\,(d-3)\,(d-2)\,(d-1)\,d\,(3\,d-4)\,(3\,d-2)\,(m_t^2-s)^2\,((3\,d-8)\,m_t^2+(d-2)\,s)}
$$

$$
+\frac{(-960+d\,(3250+d\,(-4463+d\,(2956+7\,d\,(16\,d-133)))))\,m_t^4\,s}{3\,(d-3)\,(d-2)\,(d-1)\,d\,(3\,d-4)\,(3\,d-2)\,(m_t^2-s)^2\,((3\,d-8)\,m_t^2+(d-2)\,s)}
$$

$$
+\frac{(d-2)\,(12+d\,(-263+d\,(474+d\,(56\,d-291))))\,m_t^2\,s^2}{3\,(d-3)\,(d-2)\,(d-1)\,d\,(3\,d-4)\,(3\,d-2)\,(m_t^2-s)^2\,((3\,d-8)\,m_t^2+(d-2)\,s)}
$$

$$
\left.+\frac{(d-3)\,(d-2)\,(3\,d-4)\,s^3}{3\,(d-2)\,(d-1)\,d\,(3\,d-4)\,(3\,d-2)\,(m_t^2-s)^2\,((3\,d-8)\,m_t^2+(d-2)\,s)}\right)MI_{v2,\,4}(d)
$$

$$
+4\,m_t^2\,s\left(\frac{(d-2)^2\,(72+d\,(77\,d-159))\,m_t^6+(192+d\,(165d^3-853d^2+1434d-920))\,m_t^4\,s}{3\,(d-2)\,(d-1)\,d\,(3\,d-4)\,(3\,d-2)\,(m_t^2-s)^2\,((3\,d-8)\,m_t^2+(d-2)\,s)}\right.
$$

$$
\left.+\frac{3\,(d-3)\,(d-2)\,d\,(5\,d-6)\,m_t^2\,s^2-(d-3)\,(d-2)\,d^2\,s^3}{3\,(d-2)\,(d-1)\,d\,(3\,d-4)\,(3\,d-2)\,(m_t^2-s)^2\,((3\,d-8)\,m_t^2+(d-2)\,s)}\right)MI_{v2,\,9}(d),
$$

$$
\begin{aligned}
MI_{v2,\,10}(2+d) =& 2\,s\left(\frac{((74-17\,d)\,d-72)\,m_t^4 + 2\,(d-2)\,(4\,d-5)\,m_t^2\,s + (d-2)^2\,s^2}{(d-2)\,(d-1)\,(3\,d-4)\,((3\,d-8)\,m_t^2+(d-2)\,s)\,(m_t^4-s^2)}\right)MI_{v2,\,1}(d) \\
&-\left(\frac{4\,m_t^4\,((d-4)\,m_t^2-5\,(d-2)\,s}{3\,(8-10\,d+3\,d^2)\,((3\,d-8)\,m_t^2+(d-2)\,s)}\right)MI_{v2,\,10}(d) \\
&+\left(\frac{(88+d\,(23\,d-96))\,m_t^6+(-40-11\,(d-4)\,d)\,m_t^4\,s}{3\,(d-2)\,(d-1)\,(3\,d-4)\,(m_t^2-s)\,((3\,d-8)\,m_t^2+(d-2)\,s)}\right. \\
&+\left.\frac{(d-2)\,(-30+17\,d)\,m_t^2\,s^2+3\,(d-2)^2\,s^3}{3\,(d-2)\,(d-1)\,(3\,d-4)\,(m_t^2-s)\,((3\,d-8)\,m_t^2+(d-2)\,s)}\right)MI_{v2,\,3}(d) \\
&+\left(\frac{4\,(-94+d\,(124+d\,(7\,d-52)))\,m_t^6+8\,(75+d\,(-104+(47-7\,d)\,d))\,m_t^4\,s}{3\,(d-3)\,(d-2)\,(d-1)\,(3\,d-4)\,(m_t^2-s)^2\,((3\,d-8)\,m_t^2+(d-2)\,s)}\right. \\
&+\left.\frac{-4\,(d-2)\,(3\,d-8)\,(3\,d-4)\,m_t^2\,s^2}{3\,(d-3)\,(d-2)\,(d-1)\,(3\,d-4)\,(m_t^2-s)^2\,((3\,d-8)\,m_t^2+(d-2)\,s)}\right)MI_{v2,\,4}(d) \\
&+8\,m_t^2\,s\left(\frac{(d-2)\,(d-1)\,m_t^4+2\,(26+7\,(d-4)\,d)\,m_t^2\,s}{(3\,(d-2)\,(d-1)\,(3\,d-4)\,(m_t^2-s)^2\,((3\,d-8)\,m_t^2+(d-2)\,s)}\right. \\
&+\left.\frac{(d-3)\,(d-2)\,s^2}{(3\,(d-2)\,(d-1)\,(3\,d-4)\,(m_t^2-s)^2\,((3\,d-8)\,m_t^2+(d-2)\,s)}\right)MI_{v2,\,9}(d).
\end{aligned}
$$

$$(5.14)$$

The scalar integral belonging to topology $v4$ is

$$
\begin{aligned}
I_{v4}(d,a_1,a_2,a_3,a_4,a_5,a_6) =& \int \frac{d^d k_1}{(2\pi)^d}\frac{d^d k_2}{(2\pi)^d}\frac{1}{((-k_2)^2)^{a_1}\,((k_1)^2)^{a_2}\,((q_1+k_2)^2-m_t^2)^{a_3}} \\
&\times \frac{1}{((q_2+k_1)^2)^{a_4}\,((p_1+p_2+k_1)^2-m_t^2)^{a_5}\,((q_2+k_1-k_2)^2)^{a_6}}.
\end{aligned}
$$

$$(5.15)$$

There are 9 master integrals for this topology

$$
\begin{aligned}
MI_{v4,\,1}(d) &= I_{v4}(d,0,0,1,0,1,0), & MI_{v4,\,2}(d) &= I_{v4}(d,0,1,1,0,1,0), \\
MI_{v4,\,3}(d) &= I_{v4}(d,0,1,2,0,0,1), & MI_{v4,\,4}(d) &= I_{v4}(d,0,2,1,0,0,1), \\
MI_{v4,\,5}(d) &= I_{v4}(d,0,0,2,1,0,1), & MI_{v4,\,6}(d) &= I_{v4}(d,2,0,0,0,1,1), \\
MI_{v4,\,7}(d) &= I_{v4}(d,2,1,0,0,1,1), & MI_{v4,\,8}(d) &= I_{v4}(d,1,1,1,0,1,1), \\
MI_{v4,\,9}(d) &= I_{v4}(d,2,1,1,0,1,1).
\end{aligned}
$$

$$(5.16)$$

The final formulas for the dimension shift of topology $v4$ are

$$
MI_{v4,\,1}(2+d) = \frac{(4\,m_t^4)}{d^2}\,MI_{v4,\,1}(d),
$$

$$
MI_{v4,\,2}(2+d) = \frac{m_t^2\,(m_t^2+s)}{(d-1)\,d\,s}\,MI_{v4,\,1}(d) - \frac{m_t^2\,(m_t^2-s)^2}{(d-1)\,d\,s}\,MI_{v4,\,2}(d),
$$

$$MI_{v4,\,3}(2+d) = - m_t^2 \left(\frac{(8-6\,d+d^2)\,m_t^4 + 2\,(72-76\,d+19\,d^2)\,m_t^2\,s}{3\,(64-168\,d+158\,d^2-63\,d^3+9\,d^4)\,s} \right.$$

$$\left. + \frac{(88-98\,d+25\,d^2)\,s^2}{3\,(64-168\,d+158\,d^2-63\,d^3+9\,d^4)\,s} \right) MI_{v4,\,3}(d)$$

$$-(-4+d)\,(m_t^2-s)\left(\frac{(6-5\,d+d^2)\,m_t^4 + 2\,(26-28\,d+7\,d^2)\,m_t^2\,s}{3\,(-192+568\,d-642\,d^2+347\,d^3-90\,d^4+9\,d^5)\,s} \right.$$

$$\left. + \frac{(2-3\,d+d^2)\,s^2}{3\,(-192+568\,d-642\,d^2+347\,d^3-90\,d^4+9\,d^5)\,s} \right) MI_{v4,\,4}(d),$$

$$MI_{v4,\,4}(2+d) = 2\,m_t^2 \left(\frac{(d-4)\,m_t^4 + 2\,(7\,d-16)\,m_t^2\,s + (d-4)\,s^2}{3\,(-64+104\,d-54\,d^2+9\,d^3)\,s} \right) MI_{v4,\,3}(d)$$

$$+(-4+d)\,(m_t^2-s)\left(\frac{2\,(d-3)\,m_t^4 + (+7\,d-16)\,m_t^2\,s - (-2+d)\,s^2}{3\,(192-376\,d+266\,d^2-81\,d^3+9\,d^4)\,s} \right) MI_{v4,\,4}(d),$$

$$MI_{v4,\,5}(2+d) = \frac{-16\,(2\,d-5)\,(2\,d-3)\,m_t^4}{3\,(d-2)\,(d-1)\,(3\,d-8)\,(3\,d-4)}\, MI_{v4,\,5}(d),$$

$$MI_{v4,\,6}(2+d) = \frac{-16\,(2\,d-7)\,(2\,d-5)\,m_t^4}{3\,(d-3)\,(d-2)\,(3\,d-8)\,(3\,d-4)}\, MI_{v4,\,6}(d),$$

$$MI_{v4,\,7}(2+d) = \frac{(m_t^2+s)\,(-3\,m_t^4 + d\,m_t^4 + 20\,m_t^2\,s - 6\,d\,m_t^2\,s - 3\,s^2 + d\,s^2)}{3\,(d-3)\,(d-2)\,(3\,d-8)\,s^2}\, MI_{v4,\,6}(d)$$

$$- \frac{4\,(d-4)\,(m_t^2-s)^4}{3\,(d-2)\,(-10+3\,d)\,(3\,d-8)\,s^2}\, MI_{v4,\,7}(d),$$

$$MI_{v4,\,8}(2+d) = (m_t^2+s)\left(\frac{(d-3)^2\,(d-2)\,m_t^4 - 2\,(-480+d\,(345+(d-62)\,d))\,m_t^2\,s}{4\,(d-5)\,(d-4)\,(d-3)^2\,(d-2)\,m_t^2\,(m_t^2-s)^2\,s} \right.$$

$$\left. + \frac{(d-3)^2\,(d-2)\,s^2}{4\,(d-5)\,(d-4)\,(d-3)^2\,(d-2)\,m_t^2\,(m_t^2-s)^2\,s} \right) MI_{v4,\,1}(d)$$

$$-3\left(\frac{(19+d\,(2\,d-13))\,m_t^4 + 2\,(135+d\,(-95+16\,d))\,m_t^2\,s}{(d-5)\,(d-3)\,(d-2)^2\,(m_t^2-s)^2} \right.$$

$$\left. + \frac{(19+d\,(2\,d-13))\,s^2}{(d-5)\,(d-3)\,(d-2)^2\,(m_t^2-s)^2} \right) MI_{v4,\,2}(d)$$

$$+\left(\frac{-2\,(10-3\,d)^2\,(d-3)\,(d-2)\,(3\,d-8)\,m_t^{10}+}{3\,(d-4)\,(d-3)\,(d-2)^3\,(3\,d-14)\,(3\,d-10)\,(3\,d-8)\,(m_t^2-s)^3\,s} \right.$$

$$+ \frac{4\,(-199200+d\,(306148+d\,(-183280+d\,(53191+18\,d\,(-413+22\,d)))))\,m_t^8\,s}{3\,(d-4)\,(d-3)\,(d-2)^3\,(3\,d-14)\,(3\,d-10)\,(3\,d-8)\,(m_t^2-s)^3\,s}$$

$$+ \frac{4\,(-1351584+2235188\,d-1465420\,d^2+476043\,d^3-76617\,d^4+4887\,d^5)\,m_t^6\,s^2}{3\,(d-4)\,(d-3)\,(d-2)^3\,(3\,d-14)\,(3\,d-10)\,(3\,d-8)\,(m_t^2-s)^3\,s}$$

$$+ \frac{4\,(714720+d\,(-1146668+d\,(727216-15\,d\,(15191+48\,d\,(3\,d-49)))))\,m_t^4\,s^3}{3\,(d-4)\,(d-3)\,(d-2)^3\,(3\,d-14)\,(3\,d-10)\,(3\,d-8)\,(m_t^2-s)^3\,s}$$

$$\left. + \frac{-2\,(d-4)\,(33792+d\,(-45604+d\,(22628+9\,d\,(-543+43\,d))))\,m_t^2\,s^4}{3\,(d-4)\,(d-3)\,(d-2)^3\,(3\,d-14)\,(3\,d-10)\,(3\,d-8)\,(m_t^2-s)^3\,s} \right) MI_{v4,\,3}(d)$$

$$+\left(\frac{-((10-3d)^2\,(d-4)\,(d-3)\,(d-2)\,(3d-8)\,m_t^{10})}{3\,(d-4)\,(d-3)^2\,(d-2)^3\,(3d-14)\,(3d-10)\,(3d-8)\,(m_t^2-s)^3\,s}\right.$$

$$+\frac{(1480320-2641840\,d+1928008\,d^2-735528\,d^3)\,m_t^8\,s}{3\,(d-4)\,(d-3)^2\,(d-2)^3\,(3d-14)\,(3d-10)\,(3d-8)\,(m_t^2-s)^3\,s}$$

$$+\frac{(154426\,d^4-16875\,d^5+747\,d^6)\,m_t^8\,s}{3\,(d-4)\,(d-3)^2\,(d-2)^3\,(3d-14)\,(3d-10)\,(3d-8)\,(m_t^2-s)^3\,s}$$

$$+\frac{2\,(4891776-9619128\,d+7823376\,d^2-3368990\,d^3)\,m_t^6\,s^2}{3\,(d-4)\,(d-3)^2\,(d-2)^3\,(3d-14)\,(3d-10)\,(3d-8)\,(m_t^2-s)^3\,s}$$

$$+\frac{2\,(810316\,d^4-103239\,d^5+5445\,d^6)\,m_t^6\,s^2}{3\,(d-4)\,(d-3)^2\,(d-2)^3\,(3d-14)\,(3d-10)\,(3d-8)\,(m_t^2-s)^3\,s}$$

$$+\frac{2\,(4744320-9410328\,d+7723216\,d^2-3356030\,d^3)\,m_t^4\,s^3}{3\,(d-4)\,(d-3)^2\,(d-2)^3\,(3d-14)\,(3d-10)\,(3d-8)\,(m_t^2-s)^3\,s}$$

$$+\frac{2\,(814020\,d^4-104445\,d^5+5535\,d^6)\,m_t^4\,s^3}{3\,(d-4)\,(d-3)^2\,(d-2)^3\,(3d-14)\,(3d-10)\,(3d-8)\,(m_t^2-s)^3\,s}$$

$$+\frac{(730368-1394512\,d+1097848\,d^2-456488\,d^3+105806\,d^4-12969\,d^5+657\,d^6)\,m_t^2\,s^4}{3\,(d-4)\,(d-3)^2\,(d-2)^3\,(3d-14)\,(3d-10)\,(3d-8)\,(m_t^2-s)^3\,s}$$

$$+\frac{-(d-4)^2\,(d-3)\,(d-2)\,(3d-14)\,(3d-10)\,s^5}{3\,(d-4)\,(d-3)^2\,(d-2)^3\,(3d-14)\,(3d-10)\,(3d-8)\,(m_t^2-s)^3\,s}\right)\,MI_{v4,\,4}(d)$$

$$+(m_t^2+s)\left(\frac{4\,(d-4)\,(d-3)^2\,(d-2)\,m_t^8-(d-3)\,(d-2)\,(1184+d\,(-742+113\,d))\,m_t^6\,s}{3\,(d-3)\,(d-2)^2\,(3d-10)\,(3d-8)\,s^2}\right.$$

$$+\frac{-2\,(104064+d\,(-142640+d\,(72562+d\,(-16237+1349\,d))))\,m_t^4\,s^2}{3\,(d-3)\,(d-2)^2\,(3d-10)\,(3d-8)\,s^2}$$

$$+\frac{-(d-3)\,(d-2)\,(1184+d\,(-742+113\,d))\,m_t^2\,s^3+4\,(d-4)\,(d-3)^2\,(d-2)\,s^4}{3\,(d-3)\,(d-2)^2\,(3d-10)\,(3d-8)\,s^2}\right)\,MI_{v4,\,6}(d)$$

$$+2\left(\frac{-4\,(d-4)\,(d-3)\,m_t^8+(480+d\,(-298+43\,d))\,m_t^6\,s-6\,(480+d\,(-360+67\,d))\,m_t^4\,s^2}{3\,(d-3)\,(d-2)^2\,(3d-10)\,(3d-8)\,s^2}\right.$$

$$+\frac{(480+d\,(-298+43\,d))\,m_t^2\,s^3-4\,(d-4)\,(d-3)\,s^4}{3\,(d-3)\,(d-2)^2\,(3d-10)\,(3d-8)\,s^2}\right)\,MI_{v4,\,7}(d)$$

$$+\left(\frac{-((d-4)\,(d-2)\,m_t^{10})+4\,(d-2)\,(7d-22)\,m_t^8\,s+6\,(560+d\,(-386+63\,d))\,m_t^6\,s^2}{(d-3)\,(d-2)^3\,(m_t^2-s)^2\,s}\right.$$

$$+\frac{4\,(d-2)\,(7d-22)\,m_t^4\,s^3-(d-4)\,(d-2)\,m_t^2\,s^4}{(d-3)\,(d-2)^3\,(m_t^2-s)^2\,s}\right)\,MI_{v4,\,8}(d)$$

$$+\left(\frac{4\,(2d-9)\,m_t^4\,(d-2)\,m_t^4+2\,(-40+17\,d)\,m_t^2\,s+(d-2)\,s^2}{(d-2)^3\,(12-7d+d^2)\,s}\right)\,MI_{v4,\,9}(d),$$

$$MI_{v4,\,9}(2+d)=\left(\frac{-3\,(d-2)\,(3d-11)\,(m_t^2+s)}{2\,(d-3)^2\,(20-9d+d^2)\,m_t^2\,(m_t^2-s)^2}\right)\,MI_{v4,\,1}(d)$$

$$+\left(\frac{(d-5)\,m_t^4+2\,(-61+17\,d)\,m_t^2\,s+(d-5)\,s^2}{2\,(15-8d+d^2)\,(m_t^3-m_t\,s)^2}\right)\,MI_{v4,\,2}(d)$$

$$+\Bigg(\frac{-2\left(6736+3\,d\left(-2324+d\left(852+d\left(6\,d-127\right)\right)\right)\right)m_t^6}{(d-4)\,(d-3)\,(d-2)\,(3\,d-14)\,(3\,d-10)\,(3\,d-8)\,(m_t^2-s)^3}$$

$$+\frac{-2\left(71056+d\left(-87388+d\left(39952-8049\,d+603\,d^2\right)\right)\right)m_t^4\,s}{(d-4)\,(d-3)\,(d-2)\,(3\,d-14)\,(3\,d-10)\,(3\,d-8)\,(m_t^2-s)^3}$$

$$+\frac{2\,(d-4)\left(-9908+d\left(9296+3\,d\left(-953+96\,d\right)\right)\right)m_t^2\,s^2}{(d-4)\,(d-3)\,(d-2)\,(3\,d-14)\,(3\,d-10)\,(3\,d-8)\,(m_t^2-s)^3}$$

$$+\frac{2\,(d-4)\,(d-3)\,(3\,d-14)\,(3\,d-10)\,s^3}{(d-4)\,(d-3)\,(d-2)\,(3\,d-14)\,(3\,d-10)\,(3\,d-8)\,(m_t^2-s)^3}\Bigg)\,MI_{v4,\,3}(d)$$

$$+\Bigg(\frac{-\left((d-4)\left(6736+3\,d\left(-2324+d\left(852+d\left(6\,d-127\right)\right)\right)\right)m_t^6\right)}{(d-4)\,(d-3)^2\,(d-2)\,(3\,d-14)\,(3\,d-10)\,(3\,d-8)\,(m_t^2-s)^3}$$

$$+\frac{\left(242624+d\left(-373328+d\left(228276+d\left(-69364+21\left(499-30\,d\right)d\right)\right)\right)\right)m_t^4\,s}{(d-4)\,(d-3)^2\,(d-2)\,(3\,d-14)\,(3\,d-10)\,(3\,d-8)\,(m_t^2-s)^3}$$

$$+\frac{\left(229952+d\left(-358064+d\left(221716+d\left(-68188+3\left(3469-210\,d\right)d\right)\right)\right)\right)m_t^2\,s^2}{(d-4)\,(d-3)^2\,(d-2)\,(3\,d-14)\,(3\,d-10)\,(3\,d-8)\,(m_t^2-s)^3}$$

$$+\frac{-(d-4)^2\,(2\,d-5)\,(3\,d-14)\,(3\,d-10)\,s^3}{(d-4)\,(d-3)^2\,(d-2)\,(3\,d-14)\,(3\,d-10)\,(3\,d-8)\,(m_t^2-s)^3}\Bigg)\,MI_{v4,\,4}(d)$$

$$+(m_t^2+s)\,\Bigg(\frac{(d-4)\,(d-3)\,(d-2)\,m_t^4+\left(-2592+d\left(2450+d\left(-765+79\,d\right)\right)\right)m_t^2\,s}{(d-4)\,(d-3)^2\,(d-2)\,(3\,d-8)\,(m_t^2-s)^2\,s}$$

$$+\frac{(d-4)\,(d-3)\,(d-2)\,s^2}{(d-4)\,(d-3)^2\,(d-2)\,(3\,d-8)\,(m_t^2-s)^2\,s}\Bigg)\,MI_{v4,\,6}(d)$$

$$-\Bigg(\frac{4\,(d-4)\,m_t^4+(32-11\,d)\,m_t^2\,s+(d-4)\,s^2}{\left(-240+242\,d-81\,d^2+9\,d^3\right)s}\Bigg)\,MI_{v4,\,7}(d)$$

$$-\Bigg(\frac{2\,m_t^2\,(d-2)\,m_t^4+2\left(-64+17\,d\right)m_t^2\,s+(d-2)\,s^2}{\left(6-5\,d+d^2\right)(m_t^2-s)^2}\Bigg)\,MI_{v4,\,8}(d)$$

$$+\frac{24\left(-9+2\,d\right)m_t^4}{(-4+d)\,(-3+d)\,(-2+d)}\,MI_{v4,\,9}(d). \tag{5.17}$$

At last we state the formulas for the dimension shift belonging to topology $v5$. The scalar integral belonging to topology $v5$ is

$$I_{v5}(a_1,a_2,a_3,a_4,a_5,a_6)=\int\frac{d^d k_1}{(2\pi)^d}\frac{d^d k_2}{(2\pi)^d}\frac{1}{(k_1^2)^{a_1}\,(k_2^2-m_t^2)^{a_2}\,((-p_1+k_1)^2)^{a_3}}$$

$$\times\frac{1}{((p_1-k_1)^2)^{a_4}\,((-p_1+k_1+k_2)^2-m_t^2)^{a_5}\,((-p_1-p_2+k_1)^2)^{a_6}}, \tag{5.18}$$

where as before k_1 and k_2 are the loop momenta and p_1 and p_2 are the incoming and q_1 and q_2 are the outgoing momenta. Since the 3rd and the 4th propagators are the same we get the following integral

$$I_{v5}(a_1, a_2, a_3, a_4, a_5) = \int \frac{d^d k_1}{(2\pi)^d} \frac{d^d k_2}{(2\pi)^d} \frac{1}{(k_1^2)^{a_1} (k_2^2 - m_t^2)^{a_2} ((-p_1 + k_1)^2)^{a_3}}$$

$$\times \frac{1}{((-p_1 + k_1 + k_2)^2 - m_t^2)^{a_4} ((-p_1 - p_2 + k_1)^2)^{a_5}}. \tag{5.19}$$

There are 4 master integrals for topology $v5$

$$MI_{v5,\,1}(d) = I_{v5}(d, 0, 1, 0, 1, 0, 0), \quad MI_{v5,\,2}(d) = I_{v5}(d, 1, 1, 0, 0, 1, 0),$$
$$MI_{v5,\,3}(d) = I_{v5}(d, 1, 1, -1, 1, 1, 0), \quad MI_{v5,\,4}(d) = I_{v5}(d, 1, 1, 0, 1, 1, 0). \tag{5.20}$$

The final formulas for the dimension shift of topology $v5$ are

$$MI_{v5,\,1}(2+d) = \frac{4\,m_t^4}{d^2}\, MI_{v5,\,1}(d),$$

$$MI_{v5,\,2}(2+d) = -\frac{m_t^2\,s}{(d-1)\,d}\, MI_{v5,\,2}(d),$$

$$MI_{v5,\,3}(2+d) = -2\left(\frac{16\,(-15 + 31\,d - 20\,d^2 + 4\,d^3)\,m_t^4 + (28 - 28\,d + 5\,d^2 + d^3)\,m_t^2\,s - 2\,s^2}{(3\,(d-3)\,(d-2)\,(d-1)\,(3\,d-4)\,(3\,d-2)\,s)}\right.$$
$$\left. + \frac{d\,(5 - 4\,d + d^2)\,s^2}{(3\,(d-3)\,(d-2)\,(d-1)\,(3\,d-4)\,(3\,d-2)\,s)}\right) MI_{v5,\,1}(d)$$
$$+ \frac{2\,(8\,(d-1)\,(2\,d-3)\,m_t^4 + (d^2 - 4)\,m_t^2\,s + 4\,(d-2)\,(d-1)\,s^2)}{(3\,(d-2)\,(d-1)\,(3\,d-4)\,(3\,d-2))}\, MI_{v5,\,2}(d)$$
$$+ 4\left(\frac{4\,(4\,(-5 + 2\,d)\,(-3 + 2\,d)\,m_t^6 + (-2 + d)\,(-7 + 3\,d)\,m_t^4\,s - 2\,m_t^2\,s^2)}{(3\,(d-2)^2\,(3\,d-4)\,(3\,d-2)\,s)}\right.$$
$$\left. + \frac{s^2\,((10 - 3\,d)\,d\,m_t^2 - (-2 + d)^2\,s)}{(3\,(d-2)^2\,(3\,d-4)\,(3\,d-2)\,s)}\right) MI_{v5,\,3}(d)$$
$$+ \frac{8\,(d-3)\,m_t^2\,(-24\,m_t^4 + 16\,d\,m_t^4 - 8\,m_t^2\,s + 4\,d\,m_t^2\,s - 2\,s^2 + d\,s^2)}{(3\,(d-2)^2\,(3\,d-4)\,(3\,d-2))}\, MI_{v5,\,4}(d),$$

$$MI_{v5,\,4}(2+d) = -\frac{(-40\,m_t^2 + 16\,d\,m_t^2 + 2\,s - d\,s)}{(3\,(d-3)\,(d-2)\,(3\,d-4)\,s)}\, MI_{v5,\,1}(d)$$
$$+ \frac{4\,(-m_t^2 + d\,m_t^2 + 2\,s - d\,s)}{(3\,(d-2)\,(d-1)\,(3\,d-4))}\, MI_{v5,\,2}(d)$$
$$+ \frac{2\,(8\,(d-1)\,(2\,d-5)\,m_t^4 + 4\,(d-2)\,(d-1)\,m_t^2\,s + (d-2)^2\,s^2)}{(3\,(d-2)^2\,(d-1)\,(3\,d-4)\,s)}\, MI_{v5,\,3}(d)$$
$$+ \frac{4\,(d-3)\,m_t^2\,(-4\,m_t^2 + 4\,d\,m_t^2 + 2\,s - d\,s)}{(3\,(d-2)^2\,(d-1)\,(3\,d-4))}\, MI_{v5,\,4}(d). \tag{5.21}$$

All these formulas for the shift of the dimension can be derived with the help of the Lee-method eq.(2.114). We calculated these for topologies $v3$, $v4$ and $v5$ and checked the results obtained with the previous method. We illustrate this calculation for the last topology $v5$.

For the next steps we need the kinematics

$$p_1^2 = 0, \quad p_2^2 = 0, \quad p_1 \cdot p_2 = \frac{s}{2}. \tag{5.22}$$

At first we build all scalar products between the momenta occurring in the propagators in eq. (5.19) (see eq. (2.104)). Therefore, we get the following matrix

$$\mathcal{M}_{v5} = \begin{pmatrix} k_1 \cdot k_1 & k_1 \cdot k_2 & k_1 \cdot p_1 & k_1 \cdot p_2 \\ k_2 \cdot k_1 & k_2 \cdot k_2 & k_2 \cdot p_1 & k_2 \cdot p_2 \\ p_1 \cdot k_1 & p_1 \cdot k_2 & p_1 \cdot p_1 & p_1 \cdot p_2 \\ p_2 \cdot k_1 & p_2 \cdot k_2 & p_2 \cdot p_1 & p_2 \cdot p_2 \end{pmatrix}. \tag{5.23}$$

The scalar integral in eq.(5.19) has the following five propagators

$$Pr_1 = (k_1)^2, \qquad Pr_2 = (k_2^2 - m_t^2), \quad Pr_3 = (-p_1 + k_1)^2,$$
$$Pr_4 = ((-p_1 + k_1 + k_2)^2 - m_t^2), \qquad Pr_5 = (-p_1 - p_2 + k_1)^2, \tag{5.24}$$

where Pr_n is the n-th propagator. To express all scalar products by propagators, we have to introduce two propagators additionally

$$Pr_6 = (k_1 - k_2)^2, \quad Pr_7 = (k_2 - p_2)^2. \tag{5.25}$$

Then we can express all scalar products by propagators

$$k_2^2 = m_t^2 + Pr_2, \qquad\qquad k_1.k_2 = \frac{1}{2}(m_t^2 + Pr_1 + Pr_2 - Pr_6),$$

$$k_2.p_2 = \frac{1}{2}(m_t^2 + Pr_2 - Pr_7), \qquad\qquad k_1.p_2 = \frac{1}{2}(s + Pr_3 - Pr_5),$$

$$k_1^2 = Pr_1, \qquad\qquad k_1.p_1 = \frac{1}{2}(Pr_1 - Pr_3),$$

$$k_2.p_1 = \frac{1}{2}(m_t^2 + Pr_1 + 2Pr_2 + Pr_3 - Pr_4 - Pr_6). \tag{5.26}$$

We build the determinant of the matrix $\mathcal{M}_{v5}$. We use eq.(5.26) and the kinematics in eq.(5.22) and express all scalar products in det($\mathcal{M}_{v5}$) by propagators and kinematics

$$det(\mathcal{M}_{v5}) = \frac{1}{16}\Bigg\{ s^2 \left(Pr_2^2 + (Pr_3 - Pr_4)^2 - 2\, Pr_2\,(Pr_3 + Pr_4) \right) + m_t^4 \left(4\, s\, Pr_3 + (Pr_1 - 2\, Pr_3 + Pr_5)^2 \right)$$

$$+ \Big(Pr_2\,(-3\, Pr_3 + 2\, Pr_5) - (Pr_3 - Pr_5)\,(Pr_3 - Pr_4 - Pr_6)$$

$$+ Pr_1\,(Pr_2 - Pr_3 + Pr_5 - Pr_7) + Pr_3\, Pr_7 \Big)^2$$

$$- 2\, m_t^2 \Big(2\, s^2\, Pr_3 + s\, [2\, Pr_3\, Pr_4 - Pr_3\, Pr_5 - Pr_4\, Pr_5 + Pr_2\,(-6\, Pr_3 + Pr_5)$$

$$+ Pr_1\,(-Pr_2 - 5\, Pr_3 + Pr_4 + 2\, Pr_5) + 2\, Pr_3\, Pr_6 + 2\, Pr_3\, Pr_7] + (Pr_1 - 2\, Pr_3 + Pr_5)$$

$$\times\, [Pr_2\,(3\, Pr_3 - 2\, Pr_5) + (Pr_3 - Pr_5)\,(Pr_3 - Pr_4 - Pr_6) - Pr_3\, Pr_7$$

$$+ Pr_1 \left(-Pr_2 + Pr_3 - Pr_5 + Pr_7\right)\Big]\Big) + 2\,s\,\Big(Pr_2^2\,(5\,Pr_3 - 2\,Pr_5)$$

$$+ (Pr_3 - Pr_4)\,(Pr_3 - Pr_5)\,(Pr_3 - Pr_4 - Pr_6) + Pr_3\,(-Pr_3 + Pr_4 + 2\,Pr_6)\,Pr_7$$

$$+ Pr_1\,[Pr_2^2 + Pr_3^2 + Pr_2\,(6\,Pr_3 - Pr_4 - 3\,Pr_5 - Pr_7) + Pr_4\,(Pr_5 + Pr_7)$$

$$- Pr_3\,(Pr_4 + Pr_5 + 3\,Pr_7)] + Pr_2\,[2\,Pr_3^2 + Pr_5\,(3\,Pr_4 + Pr_6)$$

$$- Pr_3\,(4\,Pr_4 + Pr_5 + 3\,(Pr_6 + Pr_7))]\Big)\Big\}. \tag{5.27}$$

As in subsec. 2.3.2, eq.(2.114), each propagator Pr_n can be interpreted as a ***lowering*** operator, for instance

$$Pr_1 \cdot I(d, a_1, a_2, a_3, a_4, a_5, a_6, a_7) = I(d, a_1 - 1, a_2, a_3, a_4, a_5, a_6, a_7)\cdot \tag{5.28}$$

To lower the dimension of an arbitrary master integral of this topology $v5$, the master integral should be multiplied with $det(\mathcal{M}_{v5})$. The resulting integrals on the right hand side are integrals with the dimension $d - 2$.

For instance, we multiply $det(\mathcal{M}_{v5})$ with $MI_{v5,3}(d) = I(d, 1, 1, -1, 1, 1, 0)$. Here, we show how the first term in the $det(\mathcal{M}_{v5})$ eq.(5.27) acts on this master integral, the other terms are calculated analogously

$$I(2 + d, 1, 1, -1, 1, 1, 0) = det(\mathcal{M}_{v5})\ I(d, 1, 1, -1, 1, 1, 0)$$

$$= \frac{1}{16}\left(s^2\left(Pr_2^2 + (Pr_3 - Pr_4)^2 - 2\,Pr_2\,(Pr_3 + Pr_4)\right) + \underbrace{\cdots}_{\text{rest of } det(\mathcal{M}_{v5})}\right)I(d, 1, 1, -1, 1, 1, 0)$$

$$= \frac{1}{16}\left(s^2\left(I(d, 1, -1, -1, 1, 1, 0, 0) - 2I(d, 1, 0, -2, 1, 1, 0, 0) - 2I(d, 1, 0, -1, 0, 1, 0, 0)\right.\right.$$

$$\left.\left. + I(d, 1, 1, -3, 1, 1, 0, 0) - 2I(d, 1, 1, -2, 0, 1, 0, 0) + I(d, 1, 1, -1, -1, 1, 0, 0)\right) + \cdots\right). \tag{5.29}$$

In the next step, we replace all occurring integrals on the right hand side by master intagrals and so we obtain the same results as in eq.(5.21) for topology $v5$.

If we use the method of Tarasov for the tensor reduction we get dots. It means we have to apply the IBP method for integrals with dots. But in the Lee-method for the dimension shift we need dots as well as scalar products, i.e., negative powers of propagators. For complicated topologies involving more parameters it is difficult to prepare such integral tables. That is the case for example for double boxes. However, the application of the Lee-method is possible for vertex diagrams.

In order to get all master integrals that occur in the vertex corrections, we have to consider topologies $Sd1$ and $Sd2$ (see fig. 5.4), too. One gets all other master integrals occurring in $Sd7$ and $Sd8$ from the other vertex topologies and $Sd1$ and $Sd2$.

The scalar integral belonging to topology $Sd1$ in fig. 5.4 is

$$I_{Sd1}(d, a_1, a_2, a_3, a_4, a_5, a_6) = \int \frac{d^d k_1}{(2\pi)^d} \frac{d^d k_2}{(2\pi)^d} \frac{1}{((k_2)^2)^{a_1}((k_1)^2)^{a_2}((-p_2+k_1)^2)^{a_3}}$$
$$\times \frac{1}{((q_1+k_2)^2 - m_t^2)^{a_4}((k_1+k_2)^2)^{a_5}((p_1-q_2+k_1+k_2)^2 - m_t^2)^{a_6}}. \tag{5.30}$$

There are 14 master integrals for this topology

$$
\begin{aligned}
MI_{Sd1,\,1}(d) &= I_{Sd1}(d,0,0,0,1,0,1), & MI_{Sd1,\,8}(d) &= I_{Sd1}(d,0,2,0,1,1,1), \\
MI_{Sd1,\,2}(d) &= I_{Sd1}(d,0,0,0,1,1,1), & MI_{Sd1,\,9}(d) &= I_{Sd1}(d,1,0,1,1,1,1), \\
MI_{Sd1,\,3}(d) &= I_{Sd1}(d,0,0,1,2,1,0), & MI_{Sd1,10}(d) &= I_{Sd1}(d,1,1,1,1,1,1), \\
MI_{Sd1,\,4}(d) &= I_{Sd1}(d,0,0,2,1,1,0), & MI_{Sd1,11}(d) &= I_{Sd1}(d,1,2,0,1,0,1), \\
MI_{Sd1,\,5}(d) &= I_{Sd1}(d,0,1,0,1,2,1), & MI_{Sd1,12}(d) &= I_{Sd1}(d,2,0,1,0,1,1), \\
MI_{Sd1,\,6}(d) &= I_{Sd1}(d,0,1,0,2,1,1), & MI_{Sd1,13}(d) &= I_{Sd1}(d,2,0,1,1,1,1), \\
MI_{Sd1,\,7}(d) &= I_{Sd1}(d,0,2,0,1,1,0), & MI_{Sd1,14}(d) &= I_{Sd1}(d,2,1,0,1,0,1).
\end{aligned}
\tag{5.31}
$$

One can find the final formulas for the dimension shift of topology $Sd1$ and $Sd2$ in app. B. The scalar integral belonging to topology $Sd2$ in fig. 5.4 is

$$I_{Sd2}(d, a_1, a_2, a_3, a_4, a_5, a_6) = \int \frac{d^d k_1}{(2\pi)^d} \frac{d^d k_2}{(2\pi)^d} \frac{1}{((k_2)^2)^{a_1}((k_1)^2)^{a_2}((-p_1+k_1)^2)^{a_3}}$$
$$\times \frac{1}{((q_2+k_2)^2)^{a_4}((k_1+k_2)^2)^{a_5}((p_2-q_1+k_1+k_2)^2)^{a_6}}. \tag{5.32}$$

There are 5 master integrals for this topology

$$
\begin{aligned}
MI_{Sd2,\,1}(d) &= I_{Sd2}(d,0,0,2,1,1,0), & MI_{Sd2,\,4}(d) &= I_{Sd2}(d,2,0,1,0,1,1), \\
MI_{Sd2,\,2}(d) &= I_{Sd2}(d,0,2,0,1,1,1), & MI_{Sd2,\,5}(d) &= I_{Sd2}(d,2,1,0,0,0,1), \\
MI_{Sd2,\,3}(d) &= I_{Sd2}(d,1,1,1,1,1,1).
\end{aligned}
\tag{5.33}
$$

For the sake of completeness we show the scalar integrals belonging to $Sd7$ and $Sd8$. The scalar integral belonging to topology $Sd7$ in fig. 5.4 is

$$I_{Sd7}(d, a_1, a_2, a_3, a_4, a_5, a_6) = \int \frac{d^d k_1}{(2\pi)^d} \frac{d^d k_2}{(2\pi)^d} \frac{1}{((k_1)^2)^{a_1}((k_2)^2)^{a_2}((p_2-k_1)^2)^{a_3}}$$
$$\times \frac{1}{((q_1-k_1)^2 - m_t^2)^{a_4}((p_1-q_2-k_2)^2 - m_t^2)^{a_5}((-p_2+k_1-k_2)^2)^{a_6}}, \tag{5.34}$$

and the scalar integral belonging to topology $Sd8$ in fig. 5.4 is

$$I_{Sd8}(d, a_1, a_2, a_3, a_4, a_5, a_6) = \int \frac{d^d k_1}{(2\pi)^d} \frac{d^d k_2}{(2\pi)^d} \frac{1}{((k_1)^2)^{a_1}((k_2)^2)^{a_2}((p_1 - k_1)^2)^{a_3}}$$
$$\times \frac{1}{((q_2 - k_1)^2)^{a_4}((-p_1 + k_1 - k_2)^2)^{a_5}((p_2 - q_1 - k_2)^2)^{a_6}}. \quad (5.35)$$

On the other hand, we can calculate all vertex diagrams with the projection method as described in subsec. 2.1.3. Using this method, we need projectors which can be the spin structures. We have two spin structures, S_1 and S_3 in eq. (5.4), for the vertex diagrams. For the application of this method we need the coefficients B_i ($i = 1, 2$), calculated in subsec. 2.1.3, eq.(2.52). Using this method for the calculation we do not have any shift in the dimension of the master integrals. We illustrate the calculation with this method for a rather simple diagram shown in fig. 5.5. This diagram is labeled in my calculation as dia65. This diagram occurs in $Amplitude_{2loops}$ for single top quark production. The reason for the choice of this diagram is that this is the simplest diagram with two spin structures and small numbers of scalar integrals after the tensor reduction. In simpler diagrams where we have only one spin structure we can not really speak of the inversion of a matrix, which is an important technical feature of this calculation. On the other hand, in more complicated diagrams, the expressions generated in the intermediate steps are so large that presenting them here would be confusing.

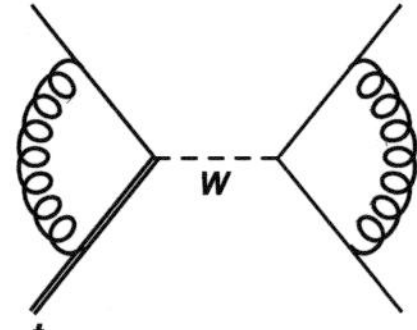

Figure 5.5.: Projections method for dia65

The scalar integral belonging to this diagram is

$$I_{dia65}(d, a_1, a_2, a_3, a_4, a_5, a_6) = \int \frac{d^d k_1}{(2\pi)^d} \frac{d^d k_2}{(2\pi)^d} \frac{1}{((k_1)^2)^{a_1}((k_2)^2)^{a_2}((p_1 - k_1)^2)^{a_3}}$$
$$\times \frac{1}{((p_2 - k_2)^2)^{a_4}((q_1 - k_2)^2 - m_t^2)^{a_5}((q_2 - k_1)^2)^{a_6}}. \quad (5.36)$$

We multiply this diagram with the conjugated spin structures of S_1 and S_3 in eq.(5.4) and sum over spins as in eq.(2.46). So we get scalar products $k_j \cdot p_i$ and $k_j \cdot q_i$ ($i, j = 1, 2$) which can be expressed by negative powers of propagators, as in subsec. 2.1.3. The scalar integral in eq.(5.36) has the following six propagators

$$Pr_1 = (k_1)^2, \qquad Pr_2 = (k_2)^2, \qquad Pr_3 = (p_1 - k_1)^2,$$
$$Pr_4 = (p_2 - k_2)^2, \qquad Pr_5 = ((q_1 - k_2)^2 - m_t^2), \qquad Pr_6 = (q_2 - k_1)^2, \tag{5.37}$$

where Pr_n is the n-th propagator. To express all scalar products by propagators we have to introduce three propagators additionally

$$Pr_7 = (p_2 + k_1)^2, \quad Pr_8 = (k_2 + q_2)^2, \quad Pr_9 = (k_1 + k_2 + p_1)^2. \tag{5.38}$$

$$\sum_{spins} \mathcal{A}_{dia65} \, \mathcal{S}^\dagger = (4\,(d-2)^2\, m_t\,(m_t^2 - 2\,s)\, B_1 + 8\,(d-2)^2\,(m_t^2 - 3\,s - t)\, B_2)\, I(d, -1, 0, 1, 1, 1, 1, 0, 0, 0)$$

$$+ (-4\,(10 - 9\,d + 2\,d^2)\, m_t\,(m_t^2 - 2\,s)\, B_1 - 8\,(d-2)\,((d-1)\, m_t^2 - (-2 + d)\,(3\,s + t))\, B_2)$$
$$\times\, I(d, -1, 1, 1, 0, 1, 1, 0, 0, 0) + 16\,(-2 + d)\,(m_t^2 + (d-4)\,(s + t))\, B_2\, I(d, -1, 1, 1, 1, 1, 1, 0, -1, 0)$$
$$-\, 16\,(d-2)\,(m_t^2\,((d-2)\,s + (d-4)\,t) - (s + t)\,((d-1)\,s + (d-3)\,t))\, B_2$$
$$\times\, I(d, -1, 1, 1, 1, 1, 1, 0, 0, 0) + (-4\,(-2 + d)^2\, m_t\,(m_t^2 - 2\,s)\, B_1 - 8\,(d-2)^2\,(m_t^2 - 3\,s - t)\, B_2)$$
$$\times\, I(d, 0, -1, 1, 1, 1, 1, 0, 0, 0) + (4\,(d-2)\, m_t\,((-8 + 3\,d)\, m_t^2 - 5\,(d-2)\,s)\, B_1 + 4\,(d-2)\,((-14$$
$$+\, 5\,d)\, m_t^2 - 4\,(d-2)\,(3\,s + t))\, B_2)\, I(d, 0, 0, 1, 0, 1, 1, 0, 0, 0) + (-4\,(d-2)^2\, m_t\,(s - t)\, B_1$$
$$-\, 8\,(d-2)\,(-2\, m_t^2 + (2 + d)\,s - (d-6)\,t)\, B_2)\, I(d, 0, 0, 1, 1, 1, 1, -1, 0, 0) + (4\,(-2 + d)\, m_t\,((2$$
$$+\, d)\, m_t^2 + (2 - 3\,d)\,s + (d-6)\,t)\, B_1 + 8\,(-2 + d)\,((2 + d)\, m_t^2 + (8 - 6\,d)\,s - 2\,d\,t)\, B_2)$$
$$\times\, I(d, 0, 0, 1, 1, 1, 1, 0, -1, 0) + (-4\,(d-2)^2\, m_t\,(m_t^2 - 2\,s)\, B_1 - 8\,(d-2)^2\,(m_t^2 - 3\,s - t)\, B_2)$$
$$\times\, I(d, 0, 0, 1, 1, 1, 1, 0, 0, -1) + (-4\,(d-2)\, m_t\,(m_t^2\,(3\,s - t) + 2\,s\,(-3\,s + (d-5)\,t))\, B_1 - 4\,(2\,(20 - 10\,d$$
$$+\, d^2)\, m_t^4 - (-152 + 100\,d - 18\,d^2 + d^3)\,s^2 - 2\,(-144 + 100\,d - 20\,d^2 + d^3)\,s\,t - (-120 + 84\,d - 18\,d^2$$
$$+\, d^3)\,t^2 + m_t^2\,((-176 + 112\,d - 20\,d^2 + d^3)\,s + (-160 + 104\,d - 20\,d^2 + d^3)\,t))\, B_2)\, I(d, 0, 0, 1, 1, 1, 1, 0, 0, 0)$$
$$+ (-4\,(d-2)\, m_t\,(2\,(-3 + d)\, m_t^2 - 3\,(d-2)\,s)\, B_1 - 4\,(d-2)\,(2\,(d-3)\, m_t^2 + (6 - 4\,d)\,s - d\,t)\, B_2)$$
$$\times\, I(d, 0, 1, 1, -1, 1, 1, 0, 0, 0) + (4\,(d-2)\, m_t\,(2\,(d-3)\, m_t^2 + (7 - 2\,d)\,s + (5 - 2\,d)\,t)\, B_1 + 8\,(-2 + d)\,((d$$
$$-\, 3)\, m_t^2 - (d-4)\,(s + t))\, B_2)\, I(d, 0, 1, 1, 0, 1, 1, -1, 0, 0) + (-4\,(d-2)\, m_t\,((2 + d)\, m_t^2 + (2 - 3\,d)\,s$$
$$+\, (d-6)\,t)\, B_1 - 4\,(d-2)\,((2 + d)\, m_t^2 + 4\,((3 - 2\,d)\,s + t - d\,t))\, B_2)\, I(d, 0, 1, 1, 0, 1, 1, 0, -1, 0)$$
$$+ (4\,(d-2)^2\, m_t\,(m_t^2 - 2\,s)\, B_1 + 4\,(d-2)^2\,(m_t^2 - 3\,s - t)\, B_2)\, I(d, 0, 1, 1, 0, 1, 1, 0, 0, -1) + (4\, m_t\,(2\, m_t^2$$
$$\times\, ((d-2)\,s + (d-4)\,t) - s\,((3\,d - 2)\,s + (2 + d)\,t))\, B_1 + 8\,((18 - 10\,d + d^2)\, m_t^4 + (s + t)\,((32 - 20\,d + 3\,d^2)$$
$$\times\, s + (28 - 18\,d + 3\,d^2)\,t) - m_t^2\,((52 - 29\,d + 4\,d^2)\,s + (56 - 31\,d + 4\,d^2)\,t))\, B_2)\, I(d, 0, 1, 1, 0, 1, 1, 0, 0, 0)$$
$$-\, 16\,(d-2)\,(m_t^2 - 2\,(s + t))\, B_2\, I(d, 0, 1, 1, 1, 1, 1, -1, -1, 0) + (8\,(d-2)\, m_t\,(m_t^2 - 2\,s)\,(s + t)\, B_1$$
$$-\, 16\,(d-2)\,(s^2 - t^2)\, B_2)\, I(d, 0, 1, 1, 1, 1, 1, -1, 0, 0) + 16\,(d-2)^2\,(s + t)\, B_2\, I(d, 0, 1, 1, 1, 1, 1, 0, -2, 0)$$
$$-\, 8\,(d-2)^2\,(s + t)\, B_2\, I(d, 0, 1, 1, 1, 1, 1, 0, -1, -1) + (4\, m_t\,(s + t)\,((2 + d)\, m_t^2 + (2 - 3\,d)\,s + (d-6)\,t)\, B_1$$
$$-\, 8\,(-4\, m_t^4 - 2\,(-2 + d)\,(s + t)\,((d-2)\,s + d\,t) + m_t^2\,((6 - 5\,d + 2\,d^2)\,s + (10 - 7\,d + 2\,d^2)\,t))\, B_2)$$
$$\times\, I(d, 0, 1, 1, 1, 1, 1, 0, -1, 0) + (-4\,(d-2)\, m_t\,(m_t^2 - 2\,s)\,(s + t)\, B_1 + 8\,(d-2)\,(-2\,(d-3)\,(s + t)^2$$

$$
\begin{aligned}
&+ m_t^2 \left((2\,d-5)\,s + (2\,d-7)\,t\right)) B_2)\, I(d,0,1,1,1,1,1,0,0,-1) + (-16\,m_t\,(s+t)\,(m_t^2\,(s-t) + s\,(-2\,s+(d \\
&-4)\,t))\, B_1 - 8\,(4\,m_t^4\,(s-t) + m_t^2\,(s+t)\,((8-10\,d+d^2)\,s + (24-10\,d+d^2)\,t) - (s+t)\,((16-10\,d+d^2)\,s^2 \\
&+ 2\,(24-12\,d+d^2)\,s\,t + (24-10\,d+d^2)\,t^2))\, B_2)\, I(d,0,1,1,1,1,1,0,0,0) + (16\,(d-2)\,m_t\,(m_t^2-s-t)\,B_1 \\
&+ 32\,(d-2)\,(m_t^2-s-t)\,B_2)\, I(d,1,-1,1,1,1,1,-1,0,0) + (8\,(d-2)\,m_t\,(m_t^2-2\,s)\,(m_t^2-s-t)\,B_1 \\
&+ 16\,(d-2)\,(m_t^4+3\,s^2+4\,s\,t+t^2-2\,m_t^2\,(2\,s+t))\, B_2)\, I(d,1,-1,1,1,1,1,0,0,0) + (-4\,(-8+2\,d+d^2) \\
&\times m_t\,(m_t^2-s-t)\,B_1 - 8\,(-4+d^2)\,(m_t^2-s-t)\,B_2)\, I(d,1,0,1,0,1,1,-1,0,0) + (-8\,(d-2)\,m_t \\
&\times (2\,m_t^2-3\,s)\,(m_t^2-s-t)\,B_1 - 8\,(d-2)\,(3\,m_t^4 - m_t^2\,(11\,s+7\,t) + 4\,(2\,s^2+3\,s\,t+t^2))\, B_2) \\
&\times I(d,1,0,1,0,1,1,0,0,0) + (-4\,(d-2)^2\,m_t\,(m_t^2-s-t)\,B_1 - 8\,(d-2)^2\,(m_t^2-s-t)\,B_2) \\
&\times I(d,1,0,1,1,1,1,-2,0,0) + (-4\,(-4+d^2)\,m_t\,(m_t^2-s-t)\,B_1 - 8\,(-4+d^2)\,(m_t^2-s-t) \\
&\times B_2)\, I(d,1,0,1,1,1,1,-1,-1,0) + (4\,(d-2)^2\,m_t\,(m_t^2-s-t)\,B_1 + 8\,(d-2)^2\,(m_t^2-s-t)\,B_2) \\
&\times I(d,1,0,1,1,1,1,-1,0,-1) + (8\,m_t\,(m_t^2-s-t)\,(d\,(s-t)+4\,t)\,B_1 + 16\,(2\,m_t^2+(d-2)\,(s-t))\,B_2) \\
&\times (m_t^2-s-t)\, I(d,1,0,1,1,1,1,-1,0,0) + (8\,(d-2)\,m_t\,(m_t^2-s-t)\,(s-t)\,B_1 + 32\,(d-2) \\
&\times (m_t^2-s-t)\,(2\,s+t)\,B_2)\, I(d,1,0,1,1,1,1,0,-1,0) - 16\,(d-2)\,(m_t^2-s-t)\,(s+t)\,B_2 \\
&\times I(d,1,0,1,1,1,1,0,0,-1) + (8\,m_t\,(m_t^2-s-t)\,(m_t^2\,(s+t) - 2\,s\,(s-(-3+d)\,t))\,B_1 + 8\,(m_t^2-s-t) \\
&\times (2\,(d-2)\,m_t^4 - (24-10\,d+d^2)\,s^2 - 2\,(24-12\,d+d^2)\,s\,t - (16-10\,d+d^2)\,t^2 + (24-12\,d+d^2)\,m_t^2 \\
&\times (s+t))\, B_2)\, I(d,1,0,1,1,1,1,0,0,0) + (4\,(d-2)\,d\,m_t\,(m_t^2-s-t)\,B_1 + 4\,(d-2)\,d\,(m_t^2-s \\
&\times I(d,1,1,1,-1,1,1,-1,0,0) - t)\,B_2) + (8\,(d-2)\,m_t\,(m_t^2-s)\,(m_t^2-s-t)\,B_1 - 4\,(d-2)\,(m_t^2-s-t) \\
&\times (-2\,m_t^2 + (2+d)\,s + d\,t)\,B_2)\, I(d,1,1,1,-1,1,1,0,0,0) + 4\,(d-2)\,m_t\,(m_t^2-s-t)\,B_1 \\
&\times I(d,1,1,1,0,1,1,-2,0,0) + (4\,(-4+d^2)\,m_t\,(m_t^2-s-t)\,B_1 + 4\,(-4+d^2)\,(m_t^2-s-t)\,B_2) \\
&\times I(d,1,1,1,0,1,1,-1,-1,0) + (-4\,(d-2)^2\,m_t\,(m_t^2-s-t)\,B_1 - 4\,(d-2)^2\,(m_t^2-s-t)\,B_2) \\
&\times I(d,1,1,1,0,1,1,-1,0,-1) + (-4\,m_t\,(m_t^2-s-t)\,((2+d)\,s + (d-2)\,t)\,B_1 - 16\,(m_t^2-s-t)\,(m_t^2 \\
&+ (d-2)\,(s+t))\,B_2)\, I(d,1,1,1,0,1,1,-1,0,0) + (-8\,(d-2)\,m_t\,(m_t^2-s-t)\,(s-t)\,B_1 \\
&- 8\,(d-2)\,(m_t^2-s-t)\,((4+d)\,s + (2+d)\,t)\,B_2)\, I(d,1,1,1,0,1,1,0,-1,0) + 8\,(2-3\,d+d^2)\,(m_t^2-s-t) \\
&\times (s+t)\,B_2\, I(d,1,1,1,0,1,1,0,0,-1) + (8\,m_t\,(-s^3+s\,t^2 - m_t^4\,(s+t) + m_t^2\,(2\,s^2+s\,t+t^2))\,B_1 - 16\,(m_t^2 \\
&- s-t)\,((d-2)\,m_t^4 + m_t^2\,((7-2\,d)\,s + (5-2\,d)\,t) + (s+t)\,((d-6)\,s + d\,t))\,B_2)\, I(d,1,1,1,0,1,1,0,0,0) \\
&+ (-8\,(d-2)\,m_t\,(m_t^2-s-t)\,(s+t)\,B_1 - 16\,(d-2)\,(m_t^2-s-t)\,(s+t)\,B_2)\, I(d,1,1,1,1,1,1,-2,0,0) \\
&+ (-4\,(2+d)\,m_t\,(m_t^2-s-t)\,(s+t)\,B_1 - 16\,(m_t^2-s-t)\,(2\,m_t^2 + (d-2)\,(s+t))\,B_2) \\
&\times I(d,1,1,1,1,1,1,-1,-1,0) + (4\,(d-2)\,m_t\,(m_t^2-s-t)\,(s+t)\,B_1 + 16\,(d-2)\,(m_t^2-s-t)\,(s+t)\,B_2) \\
&\times I(d,1,1,1,1,1,1,-1,0,-1) + (16\,m_t\,(m_t^2-s-t)\,(s^2-t^2)\,B_1 + 32\,(m_t^2-s-t)\,(s^2-t^2)\,B_2) \\
&\times I(d,1,1,1,1,1,1,-1,0,0) - 4\,(-12+4\,d+d^2)\,(m_t^2-s-t)\,(s+t)\,B_2\, I(d,1,1,1,1,1,1,0,-2,0) \\
&+ 8\,(d-2)\,d\,(m_t^2-s-t)\,(s+t)\,B_2\, I(d,1,1,1,1,1,1,0,-1,-1) + (8\,m_t\,(m_t^2-s-t)\,(s^2-t^2)\,B_1 \\
&- 16\,(m_t^4\,(s-t) + 4\,(s-t)\,(s+t)^2 - 5\,m_t^2\,(s^2-t^2))\,B_2)\, I(d,1,1,1,1,1,1,0,-1,0) - 4\,(d-2)^2\,(m_t^2-s-t) \\
&\times (s+t)\,B_2\, I(d,1,1,1,1,1,1,0,0,-2) + 16\,(m_t^4\,(s-t) + 2\,(s-t)\,(s+t)^2 - 3\,m_t^2\,(s^2-t^2))\,B_2 \\
&\times I(d,1,1,1,1,1,1,0,0,-1) + (32\,m_t\,s\,(m_t^2-s-t)\,t\,(s+t)\,B_1 + 16\,(m_t^2-s-t)\,(s+t)\,(-((d-2)\,s^2) \\
&- 2\,(d-4)\,s\,t - (d-2)\,t^2 + (d-2)\,m_t^2\,(s+t))\,B_2)\, I(d,1,1,1,1,1,1,0,0,0),
\end{aligned}
$$

$$\tag{5.39}$$

where B_1 and B_2 are the coefficients in eq.(2.52) and eq.(2.53). t and s are the Mandelstam variables

$$p_1 \cdot p_2 = \frac{s}{2}, \qquad p_2 \cdot q_2 = -\frac{t}{2}, \tag{5.40}$$

m_t the top mass and d the dimension.

As we can see, after the application of the projection method we get scalar integrals with no shift in the dimension and no positive powers of the propagators but negative powers. In the next step we replace all scalar integrals in eq.(5.39) by master integrals. We use Reduze2 [25] to get the master integrals for this diagram. There are two master integrals for this diagram

$$MI_{dia65,\,1}(d) = I(d,0,0,1,0,1,1,0,0,0), \quad MI_{dia65,\,2}(d) = I(d,0,0,1,1,1,1,0,0,0),$$

$$\sum_{spins} A_{dia65}\, S^\dagger = \mathbf{B}_1\left(\left[-4\left(-32+30\,d-9\,d^2+d^3\right)s\,t\left(2\,(d-2)\,m_t^2+(18-9\,d+d^2)\,(s+t)\right)MI_{dia65,\,1}(d)\right]\right.$$

$$\left/\left((d-4)^2\,(d-3)\,m_t\,(s+t)\right)+\left[16\left(-32+30\,d-9\,d^2+d^3\right)m_t\,s\,(m_t^2-s-t)\,t\,MI_{dia65,\,2}(d)\right]\right/\left((d-4)^2\,(s+t)\right)\right)$$

$$+\,\mathbf{B}_2\left(\left[-4\left(-32+30\,d-9\,d^2+d^3\right)\left((d-2)^2\,m_t^4\,(s+t)+(s+t)\left((d-2)\,s^2+2\,(d-4)\,s\,t+(d-2)\,t^2\right)\right.\right.\right.$$

$$\left.\left.-\,m_t^2\left((2-3\,d+d^2)\,s^2+4\,(2\,d-7)\,s\,t+(2-3\,d+d^2)\,t^2\right)\right)MI_{dia65,\,1}(d)\right]\left/\left((d-4)^2\,(d-3)\,m_t^2\,(s+t)\right)\right.$$

$$+\left[4\,(16-7\,d+d^2)\,(m_t^2-s-t)\left(2\,(d-2)^2\,m_t^2\,(s+t)-(16-7\,d+d^2)\,((d-2)\,s^2\right.\right.$$

$$\left.\left.\left.+\,2\,(d-4)\,s\,t+(d-2)\,t^2\right)\right)MI_{dia65,\,2}(d)\right]\left/\left((d-4)^2\,(s+t)\right)\right). \tag{5.41}$$

After all, we obtain the final result for this diagram after replacing B_1 and B_2 from eq.(2.52) and eq.(2.53). Note that S_2 in eq.(2.52) and eq.(2.53) is equal to S_3 in eq.(5.4).

$$A_{dia65} = \left(\frac{(d-5)\,(d-2)\,(16-7\,d+d^2)\,S_1}{(d-4)\,(d-3)\,m_t\,(s+t)}+\frac{(d-2)\,(16-7\,d+d^2)\,(-2\,m_t^2+d\,m_t^2-s-t)\,S_3}{((d-4)^2\,(d-3)\,m_t^2\,(s+t))}\right)MI_{dia65,\,1}(d)+$$

$$\left(\frac{-2\,(d-5)\,(16-7\,d+d^2)\,m_t\,S_1}{(d-4)\,(s+t)}+\frac{(16-7\,d+d^2)\,(4\,m_t^2-2\,d\,m_t^2+16\,s-7\,d\,s+d^2\,s+16\,t-7\,d\,t+d^2\,t)\,S_3}{(d-4)^2\,(s+t)}\right)MI_{dia65,\,2}(d), \tag{5.42}$$

where S_1 and S_3 are defined in eq.(5.4). Based on this rather simple example we have seen that the intermediate expression in eq.(5.39) is getting long, compared to the final result in eq.(5.42). The intermediate expression for this diagram calculated with the other method, the Tarasov-method, is longer than in eq.(5.39).

5.2.2. Results

We present the final result of the vertex contributions in app. C. We computed the vertex contributions in two independent calculations and with two different methods. The final results reached by the two methods and the two calculations are equal with each other. But nevertheless, we want to compare the results with literature, too. We can find two kinds of contributions in our results, the contribution of corrections to the light quarks side which can be seen in fig. 5.6

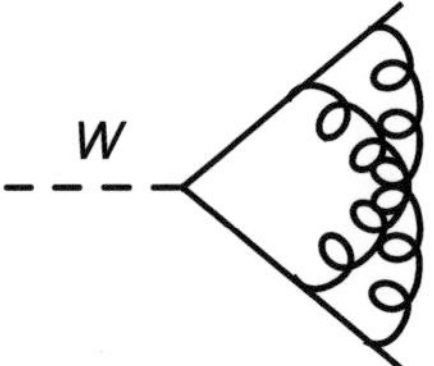

Figure 5.6.: Light quark form factor

and the contribution of corrections to the Light-to-Heavy quark side depicted in fig. 5.7

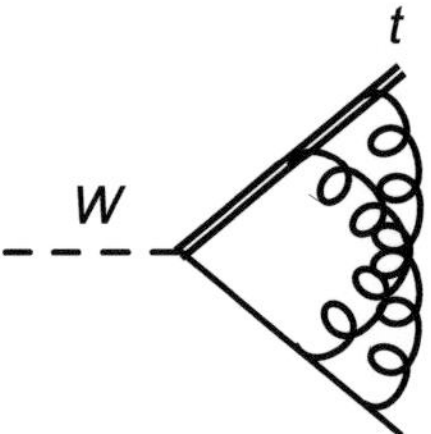

Figure 5.7.: Light-to-Heavy quark decay

Our result for the light quark form factor is consistent with [100], and the heavy quark side (shown in fig. 5.7) is consistent with [101] which is independently confirmed by other groups [102–104]. We have the analytic expressions for the master integrals from [105]. However, the analytic expressions for the master integrals can be found in [106–111], too.

5.2.2.1. C code

In order to evaluate our expressions numerically we wrote a C code which can be used as a static library. In the final result of the vertex contributions there are two spin structures, S_1 and S_3, and four colour coefficients, $\mathcal{C}_1 = C_F^2 N^2$, $\mathcal{C}_2 = C_F C_A N^2$, $\mathcal{C}_3 = I_2(R) C_F N_H N^2$ and $\mathcal{C}_4 = I_2(R) C_F N_L N^2$ as represented in tab. 5.1, where N_H and N_L are the labels for the heavy-quark loop and light-quark loops. That motivates us to factorize the final result to coefficients of the spin structures and the colour factors. The final expression belonging to vertex contributions can be considered as a matrix

$$vertices = \sum_{i=1}^{2} \sum_{j=1}^{4} S_i \, c_{ij} \, \mathcal{C}_j. \tag{5.43}$$

In the next step we consider the Laurent expansion of each coefficient $c_{i,j}$ for $i \in \{1,2\}$ and $j \in \{1,\cdots,4\}$ in this matrix from ϵ^{-4} to ϵ^0

$$vertices = \sum_{i=1}^{2} \sum_{j=1}^{4} \sum_{k=-4}^{0} S_i \, c_{ijk} \, C_j \, \epsilon^k. \tag{5.44}$$

Every entry can be represented clearly by a 3 dimensional array $c[spin][colour][\epsilon]$.

The Harmonic polylogarithms (HPLs) occurring in our final result can be evaluated by using CHAPLIN (Complex HArmonic PolyLogarithms In fortraN) library [112]. We make a library, named *"vertlib.a"*, available which contains all expressions needed for the calculation of the coefficients of the matrix in eq.(5.44). Every coefficient, $c[spin][colour][\epsilon]$, depends on the Mandelstam variable t and the mass of the top quark. In order to know which values we can take for the kinematical variables t, we consider

$$t < 0, \quad u < 0 \quad \text{and} \quad m_t^2 < s < S_{\text{hadronic}},$$

and

$$m_t^2 = t + s + u. \tag{5.45}$$

Thus, we can estimate the following region for t

$$m_t^2 - S_{\text{hadronic}} < t < 0. \tag{5.46}$$

Using this library *"vertlib.a"* we get a numerical value for each coefficient, c_{ijk} from eq.(5.44), up to a global factor

$$C_{global,vertices} = \frac{\alpha_e \alpha_s V_{tb} V_{ud}^*}{8 \, \sin^2 \theta_W (t - m_w^2)} \left(\frac{\alpha_s}{4\pi}\right), \tag{5.47}$$

where α_s is the strong coupling constant and α_e is the fine-structure constant. V_{tb}, V_{ud}^*, θ_W and m_w are defined as in eq. (3.6).

One can download the library *"vertlib.a"* and the final result as a Mathematica file from: `http://people.physik.hu-berlin.de/~assadsol`.

5.3. Double box contributions

There are in total 9 topologies, which represent the double box diagrams, three planar double boxes (shown in fig. 5.2) and six non-planar double boxes (shown in fig. 5.3).[3]

Two diagrams, which differ in the kinematic variables, belong to each topology. We can use the integration tables which we get from the application of the IBP method to one diagram for the other one by matching the kinematic variables. Two methods are presented in chapter 2 for the tensor reduction: first, the Tarasov-method in subsec. 2.1.2 and second, the projection method in subsec. 2.1.3. Depending on the chosen method for the tensor

---

[3]The concept of topology was defined in section 2.3 and used in the last section 5.2, too.

reduction, we will have a different starting point for the application of the IBP method. Using the first method described in subsec. 2.1.2, we will have integrals with positive powers of the propagators, the so-called dots (for the definition of dots see eq.(2.28)) and a shift in the dimension of the integrals. In this case, we obtain a maximum of 8 dots because we have two loops and tensors of the fourth rank and a maximum shift in the dimension of integrals, d, up to 8. That makes the application of the IBP method (reduction to master integrals) extremely difficult. We discuss the reduction to master integrals in the next subsection. So, we favor the application of the projection method for the tensor reduction. By the application of the projection method we get scalar products between the external momenta and the loop momenta. We can express these scalar products by negative powers of the propagators and we get nor shift in the dimension neither dots as explained in subsec. 2.1.3. To apply this method we need projectors (see eq.(2.41)). The spin structures can be used as projectors. We used the spin structures shown in eq.(5.4) as projectors. We need the tensor coefficients for the application of the projection method, as described in subsec. 2.1.3, which are calculated in app. E.

In the calculation of double boxes, the five variables s, t, m_t, m_w and d occur, where t and s are the Mandelstam variables, m_t the mass of the top quark, m_w the mass of the W boson and d the space-time dimension. The calculation of such diagrams is equivalent to the treatment of multivariate polynomials. In this sense the number of variables determines the complexity of the calculation. Especially the step, reduction to master integrals, turns out to be difficult. There is some software for this aim which is public like Air [46] in Maple, Fire [47] in Mathematica, Reduze1 [48] and Reduze2 [25] in $C++$. We mainly used *Reduze2* [25] for the reduction to master integrals. The advantage of *Reduze2* is that it is full parallelized and uses Fermat [113] to treat the polynomial expressions, which is more efficient compared to other computer algebra systems [25, 114]. However, these properties are not sufficient to solve the problem of the reduction of non-planar double boxes by brute force application of standard software. In the next subsection, we illustrate the intermediate steps which we have done to get the reduction tables for the non-planar double boxes.

5.3.1. Reduction to master integrals

By using the projection method from subsec. 2.1.3 we get scalar integrals with four scalar products after tensor reduction. Each scalar product can be represented by a negative power of a propagator. In these topologies, planar and non-planar double boxes, seven propagators occur. However, we have to introduce two additional propagators in order to be able to express all scalar products by propagators. Therefore, we have scalar Feynman integrals with the following structure

$$I(a_1, \cdots, a_9) = \int \int d^d k_1 \, d^d k_2 \, \frac{1}{Q_1^{a_1} \cdots Q_9^{a_9}}, \quad -4 \leq a_i \leq 1 \, \forall i \in \{1, \cdots, 9\}, \quad (5.48)$$

where $d = 4 - 2\epsilon$ and $Q_i, i = 1, \cdots, 9$ are the propagators. The sum over all negative a_i is greater than or equal to -4, i.e., we have a maximum of 4 scalar products in our calculation. In this section we discuss at first how the reduction works in the sense of an implemented Laporta's algorithm [24]. Then we talk about the main points and features of Reduze2 [25]. Finally, we combine the knowledge from these two parts to organize the reduction.

5.3.1.1. Reduction process

We consider the following scalar Feynman integral

$$I(a_1, \cdots, a_n) = \int \cdots \int d^d k_1 \cdots d^d k_h \, \frac{1}{Q_1^{a_1} \cdots Q_n^{a_n}}, \tag{5.49}$$

where $d = 4 - 2\epsilon$ and in the denominator Q_i , $i = 1, \cdots, n$ are the propagators. The n-dimensional integer space in which a Feynman integral lives can be divided into subsets, which are called sectors. For instance, if we have six propagators we are in the 6th sector, but we will have more possibilities to place the propagators because we have seven propagators indeed[4], i.e., we have to put six propagators to seven places.
For example

$$I(0, a_2, a_3, a_4, a_5, a_6, a_7, a_8, a_9),$$

$$\vdots$$

$$I(a_1, a_2, a_3, a_4, a_5, a_6, 0, a_8, a_9) . \tag{5.50}$$

They are called sub-sectors[5] and can be numerated in this way

$$\text{number of sub-sector}\Big(I(a_1, a_2, a_3, a_4, a_5, a_6, a_7, a_8, a_9) \Big) = \sum_{i=1}^{9} \Theta(a_i - \epsilon) 2^i, \tag{5.51}$$

where Θ is the Heaviside step function and $0 < \epsilon < \frac{1}{2}$. The sector number and the sub-sector number can be used to identify each scalar integral, for instance, $I(0, 1, 2, 1, 1, 1, 1, -2, 0)$ lives in sector 6 and sub-sector 126, which is represented as $I_{6_126}(0, 1, 2, 1, 1, 1, 1, -2, 0)$.

We can represent sectors in a graphical way. In general it might be not practical because the representation of a dimension greater than three is difficult. However, it could be helpful in some cases when we discuss the dependency of two sectors (see fig. 5.8).

Each sector has a corner. The corner is the point with $a_i = 1$ if we have dots, i.e., $a_i > 1$ or the point with $a_i = 0$ in case of scalar products, i.e. $a_i < 0$. The aim of a reduction is to move a_i into the corner as illustrated in fig. 5.8. There are many possibilities to choose a route from a starting point into the corner. The arrows in fig. 5.8 were represented in subsec. 2.2.2 by *increasing* and *lowering* operators. That is a graphical interpretation of solving IBP relations. The goal of solving IBP relations is to express a given complicated integral by a linear combination of simpler integrals. Simpler integrals exist normally in lower sectors. In other words, reduction means to find relations between integrals in different sectors. The reduction problem is recursive, i.e., if one wants to solve the reduction problem for a

---

[4]We do not consider both propagators Q_8 and Q_9 because they do not really occur in the integral. They are just used to represent irreducible scalar products.

[5]One can find in the literature, for instance in [48], another definition for the sub-sector but here this definition is used because of the explanation of our method.

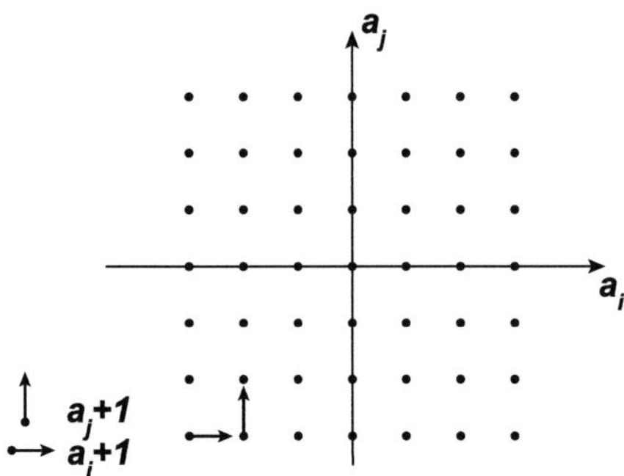

Figure 5.8.: Representation of sectors in n-dimensional integer space. $i, j \in \{1, \cdots, n\}$ in $I(a_1, \cdots, a_n)$ eq.(5.49)

given integral in a certain sector one has to solve it for all other sectors which lie below that sector. Standard implementations of the Laporta's algorithm [24] generate seed integrals (seeds) and insert them in the IBP identities. Then they try to solve these identities for more complicated integrals. We can consider seeds as starting points. We empirically know that the choice of seeds might enable us to improve the time of the calculation. Before discussing this in details we give a brief introduction to using Reduze2 [25].

5.3.1.2. A brief insight into Reduze2

Reduze2 [25] uses a subdirectory, named *config*, containing normally three files:

```
config/kinematics.yaml
config/integralfamilies.yaml
config/global.yaml
```

where the file *kinematics.yaml* defines the kinematics and *integralfamilies.yaml* contains the propagators occurring in the integral. The file *global.yaml* collects other global configuration data. For instance, if we want to use Fermat [113] instead of GiNaC [115] we have to put this information in that file. In order to begin with the reduction we need a file which contains information concerning the number of dots or scalar products, which sector we want to reduce, how we want to reduce, recursively or not, etc. This file is called *jobs_reduction.yaml*. The file typically looks like this

```
jobs:
- setup_sector_mappings: {}
- reduce_sectors:
    conditional: true
    sector_selection:
      select_recursively:
```

```
        - [topo, 127]
  alternative_input_directory: "reductions_s1"

  identities:
    ibp:
      - { r: [t, 7], s: [2, 2]}
    lorentz:
      - { r: [t, t], s: [2, 2]}
    sector_symmetries:
      - { r: [t, t], s: [2, 2]}
  reduzer_options:
   use_database: true
   use_transactions: true
   cache_size: 536870912
```

$$(5.52)$$

In this job-file we can see that the reduction method is chosen recursively. The highest sector where we begin is sector 7, sub-sector 127, i.e., with seven propagators in eq.(5.51). In *identities* we can determine which identities we use and whether we need dots, $-\{\mathbf{r} : [\mathbf{t}, \mathbf{7}], s : [\cdots]\}$, where t is the number of propagators and for the sub-sectors we have $(7 - t)$ dots or scalar products, $-\{r : [\cdots], \mathbf{s} : [\mathbf{0}, \mathbf{2}]\}$, where we have up to two scalar products for the reduction. The option "*use_database : true*" in "*reduzer_options*" allows us to create a database on the hard disk drive. The whole reduction runs in RAM if one sets it as follows:

```
      reduzer_options:
       use_database: false
#        use_transactions: true   # comment out
#        cache_size: 536870912
```

where "*use_transactions : true*" causes that the reduction runs after an interruption again from the interrupted point and not from the beginning and "*cache_size :*" determines the capacity of the hard disk which will be reserved and can be maximum about 4 GB.

alternative_input_directory can be used if we do the reduction in many steps or insert additional identities from other sources to the reduction of a selection of sectors. We can organize the reduction in two or more steps. For example, we run the reduction at first for

```
      identities:
        ibp:
          - { r: [t, 7], s: [0, 1]}
```

and save the results as *reductions_s1*. In the second run we do the reduction for the following setup:

```
      alternative_input_directory: "reductions_s1"
      identities:
        ibp:
          - { r: [t, 7], s: [2, 2]}
```

and save the results as *reductions*. The composition of two files should contain the reduction which we get with the following setup:

```
identities:
  ibp:
    - { r: [t, 7], s: [0, 2]}
```

This feature can help us to do the reduction in many steps in case of a complicated topology.

5.3.1.3. Organization of the reduction

After application of the projection method, as shown in subsec. 2.1.3, we get scalar integrals which have up to four scalar products. In other words, we have to do a reduction as follows:

```
identities:
  ibp:
    - { r: [t, 7], s: [0, 4]}
```

However, for each diagram we have three scalar integrals with four scalar products and in the highest sector, namely sector 7_127, integrals like

$$I(1,1,1,1,1,1,1,-4,0),$$
$$I(1,1,1,1,1,1,1,-3,-1),$$
$$\vdots$$

(5.53)

To do this reduction in one run is very difficult for the planar double boxes in fig. 5.2 and almost impossible for the non-planar double boxes in fig. 5.3[6]. So we have to do the reduction in many steps. The first idea to do that can look like this first run

```
 jobs:
- setup_sector_mappings: {}
- reduce_sectors:
    conditional: true
    sector_selection:
      select_recursively:
        - [topo7, 127]
    identities:
      ibp:
        - { r: [t, 7], s: [0, 3]}
      lorentz:
        - { r: [t, t], s: [0, 3]}
      sector_symmetries:
        - { r: [t, t], s: [0, 0]}
```

(5.54)

---

[6]Here, "almost impossible" means that we need very huge computer resources, especially RAM, and a very long run time.

this can be saved as *reductions_s3*. In the second run we can set the job file as follows:

```
jobs:
- setup_sector_mappings: {}
- reduce_sectors:
    conditional: true
    sector_selection:
      select_recursively:
        - [topo7, 127]
    alternative_input_directory: "reductions_s3"
    identities:
      ibp:
        - { r: [t, t], s: [4, 4]}
#       lorentz:                       # comment out
#         - { r: [t, t], s: [2, 2]}
#       sector_symmetries:
#         - { r: [t, t], s: [2, 2]}
```

$$(5.55)$$

One might come to the idea that we can continue doing the reduction in that way. We reduced the topology 7, *d7*, in fig. 5.2 in that way and recognized that it is very time consuming. This means , it will be more difficult for the non-planar topologies. The next idea could be to do the reduction for non-planar double boxes, fig. 5.3, in more steps. Before we begin with the reduction, we have to consider:

- The reduction should be done recursively.

- By the partition of the reduction we lose some relations between sectors, which can be important to enable us to find all master integrals.

- We need one reduction table in which we can definitely find all master integrals, at least in the highest sector.

We start with the first reduction:

```
jobs:
- setup_sector_mappings: {}
- reduce_sectors:
    conditional: true
    sector_selection:
      select_recursively:
        - [topo2, 127]
    identities:
      ibp:
        - { r: [t, 8], s: [0, 1]}
      lorentz:
        - { r: [t, t], s: [0, 1]}
      sector_symmetries:
        - { r: [t, t], s: [0, 1]}
```

save the result as *reductions_s1*, and start with the second run:

```
 jobs:
 - setup_sector_mappings: {}
 - reduce_sectors:
     conditional: true
     sector_selection:
       select_recursively:
         - [topo2, 127]
     alternative_input_directory: "reductions_s1"
     identities:
       ibp:
         - { r: [t, t], s: [2, 2]}
#        lorentz:                      # comment out
#          - { r: [t, t], s: [0, 1]}
#        sector_symmetries:
#          - { r: [t, t], s: [0, 1]}
```

save the result as *reductions_s2* and begin with the next reduction:

```
 jobs:
 - setup_sector_mappings: {}
 - reduce_sectors:
     conditional: true
     sector_selection:
       select_recursively:
         - [topo2, 127]
     alternative_input_directory: "reductions_s2"
     identities:
       ibp:
         - { r: [t, t], s: [3, 3]}
#        lorentz:                      # comment out
#          - { r: [t, t], s: [0, 1]}
#        sector_symmetries:
#          - { r: [t, t], s: [0, 1]}
```

and for the last step, $-\{\mathbf{r} : [\mathbf{t}, \mathbf{t}],\ s : [4, 4]\}$ it can be done in the same way. But if we use all these files *reductions_s1*, *reductions_s2*, *reductions_s3* and *reductions_s4* together to replace our scalar integrals by master integrals we do not find all master integrals immediately. However, these integrals which are not master integrals and occur on the right hand side, exist normally in the lower sectors, lower than the sixth sector. In order to get all master integrals we have to repeat the reduction for these sectors. Finally, we have 5 reduction tables. This consideration seems to work. But it only can be applied to topology 2, $d2$ in fig. 5.3, and topology 4, $d4$ in fig. 5.3. We can not apply the last step of this strategy to other topologies, i.e., $-\{\mathbf{r} : [\mathbf{t}, \mathbf{t}],\ s : [4, 4]\}$ does not work for topologies 1,3,5 and 6 as good as for topology 2 and topology 4 in fig. 5.3. The main reason is that we have the mass of the top quark in the loops. We know that our scalar integrals with four scalar products occur only in sector 7_127. Perhaps we can try to reduce just this sector. However, we have seen in subsec. 5.3.1.1 that the reduction should be done recursively. If we try to do this reduction just for this sector and not recursively, we get integrals on the right hand side which are just as complicated as the integral on the left hand side. For instance

$$I(1,1,1,1,1,1,1,-4,0) \to I(1,1,1,1,1,1,1,-2,-2)\cdots \tag{5.56}$$

So we have to do the reduction recursively but how can we do it if we are not able to do the reduction for all sectors lying lower than 7_127, fig. 5.9

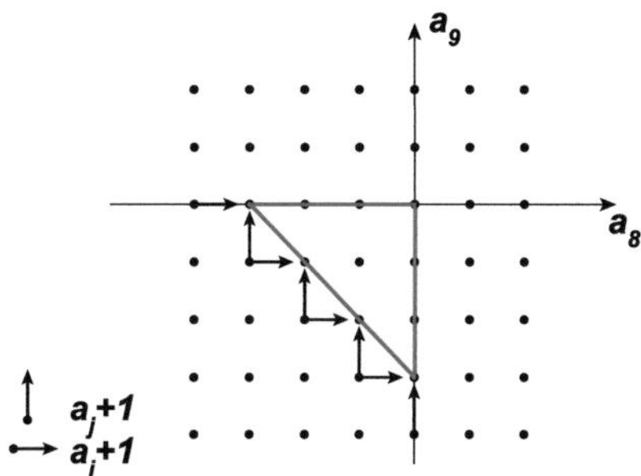

Figure 5.9.: The reduction problem of the forth scalar product. Reduction tables exist for integrals being within the green frame

As we can see in fig. 5.9 the problem is that we have reduction tables up to three scalar products and we have just some scalar integrals with four scalar   products in the last sector 7_127 [7]. We tried at first the idea to copy the folder *reductions_s3* to *reductions* and remove the last reduction file containing the reduction of 7_127 which was created by

```
jobs:
- setup_sector_mappings: {}
- reduce_sectors:
    conditional: true
    sector_selection:
      select_recursively:
        - [topo2, 127]
    alternative_input_directory: "reductions_s2"
    identities:
      ibp:
        - { r: [t, t], s: [3, 3]}
```
$$\tag{5.57}$$

Then we changed the file *jobs_reduction.yaml* as follows

---

[7]We have a maximum of three integrals with four scalar products per diagram. All of them exist in 7_127.

```
jobs:
- setup_sector_mappings: {}
- reduce_sectors:
     conditional: true
     sector_selection:
       select_recursively:
         - [topo2, 127]
     alternative_input_directory: "reductions_s3"
     identities:
       ibp:
         - { r: [t, t], s: [4, 4]}
```
$$\tag{5.58}$$

and started the reduction. That means all lower sectors were reduced just up to three scalar products and we tried just to reduce the last sector for $-\{\mathbf{r}:[\mathbf{t},\mathbf{t}],\ s:[4,4]\}$. But the problem is that the run time of this strategy is extremely long. This was running for topologies 1,3 and 6 for longer than three months and we have not gotten any result so far. So we have to build a bridge between the sub-sectors of sector six and sub-sector 7_127. That means we do the last step but we do not begin directly with sub-sector 7_127. We not only remove sub-sector 7_127 but all sixth sub-sectors. In order to get a connection between the sub-sector 7_127 and the sub-sectors of sector six we have to run the reduction with one dot in seeds at last. If we leave this task to the standard software they generate all possible seeds for each sub-sector. In this case, it makes the reduction very difficult if we take all automatic generated seeds. The idea is to take just selected seeds. But what does "selected seeds" mean and how did we select these seeds? We proceeded as follows:

- Consider the scalar integrals with four scalar products, $I(1,1,1,1,1,1,1,-a_8,-a_9)$, $|a_8|+|a_9|=4$, occurring in the problem as seeds.

- Determine all IBP relations, in which these integrals occur.

- Solve each IBP relation for such integrals $I(1,1,1,1,1,1,1,-a_8,-a_9)$ with four scalar products.

- Consider all other integrals with four scalar products and one dot in the sub-sectors with six propagators as new seeds.

To demonstrate this algorithm we apply it to topology 3, $d3$ in fig. 5.3. The scalar integral belonging to topology 3 is

$$I(d,a_1,a_2,a_3,a_4,a_5,a_6,a_7,a_8,a_9) = \int \int \frac{d^d k_1 d^d k_2}{(k_1^2)^{a_1}(k_2^2)^{a_2}((p_1-k_1)^2)^{a_3}((p_2+k_1)^2)^{a_4}} \times$$

$$\frac{1}{((q_1+k_2)^2-m_t^2)^{a_5}((p_1-k_1+k_2)^2)^{a_6}((p_1-q_2-k_1+k_2)^2-m_w^2)^{a_7}((k_1-q_2)^2)^{a_8}((k_2-p_2)^2)^{a_9}}.$$

$$\tag{5.59}$$

There are five scalar integrals with four scalar products

$$I(1,1,1,1,1,1,1,-4,0), \qquad I(1,1,1,1,1,1,1,-3,-1),$$
$$I(1,1,1,1,1,1,1,-2,-2), \qquad I(1,1,1,1,1,1,1,-1,-3),$$
$$I(1,1,1,1,1,1,1,0,-4). \tag{5.60}$$

In the next step we write all IBP relations

$$1)IBP[k_1,k_1]: -2a_1 - a_3 - a_4 - a_7 - a_8 + d - a_3\mathbf{1}^-\mathbf{3}^+ - a_4\mathbf{1}^-\mathbf{4}^+$$
$$+ a_6(m_t^2 - m_w^2 - s - t - \mathbf{2}^- + \mathbf{5}^- - \mathbf{7}^- - \mathbf{8}^- + \mathbf{9}^-)\mathbf{6}^+$$
$$+ a_7(m_t^2 - m_w^2 - s - t - \mathbf{1}^- - \mathbf{2}^- + \mathbf{5}^- + \mathbf{9}^-)\mathbf{7}^+ - a_8\mathbf{1}^-\mathbf{8}^+ = 0$$

$$2)IBP[k_1,p_1]: -a_1 - a_3 + a_6 + a_7 - a_1(m_t^2 - m_w^2 - s - t - \mathbf{2}^- + \mathbf{3}^- + \mathbf{5}^- - \mathbf{7}^- - \mathbf{8}^- + \mathbf{9}^-)\mathbf{1}^+$$
$$- a_3(\mathbf{2}^- - \mathbf{6}^-)\mathbf{3}^+ - a_4(m_t^2 - m_w^2 - s - t + \mathbf{1}^- + \mathbf{3}^- + \mathbf{5}^- - \mathbf{7}^- - \mathbf{8}^-)\mathbf{4}^+$$
$$+ a_6(\mathbf{2}^- - \mathbf{3}^-)\mathbf{6}^+ - a_7(m_t^2 - m_w^2 - s - t + \mathbf{1}^- - \mathbf{2}^- + \mathbf{3}^- - \mathbf{8}^-)\mathbf{7}^+$$
$$+ a_8(\mathbf{2}^- - \mathbf{3}^- - \mathbf{5}^- + \mathbf{6}^- - \mathbf{9}^-)\mathbf{8}^+ = 0$$

$$3)IBP[k_1,p_2]: -a_1 + a_3 + a_6 - a_7 + a_1\mathbf{3}^-\mathbf{1}^+ - a_3\mathbf{1}^-\mathbf{3}^+ - a_4(s + \mathbf{1}^- - \mathbf{3}^-)\mathbf{4}^+$$
$$+ a_6(m_t^2 - m_w^2 - s - t - 2\,\mathbf{2}^- + \mathbf{3}^- + \mathbf{5}^- - \mathbf{7}^- - \mathbf{8}^- + \mathbf{9}^-)\mathbf{6}^+$$
$$+ a_7(2m_t^2 - m_w^2 - 2s - 2t - 2\,\mathbf{2}^- + \mathbf{3}^- + \mathbf{5}^- + \mathbf{6}^- - \mathbf{8}^- + \mathbf{9}^-)\mathbf{7}^+$$
$$- a_8(m_t^2 - s - t + \mathbf{1}^- - \mathbf{3}^-)\mathbf{8}^+ = 0$$

$$4)IBP[k_1,q_2]: a_1 - a_4 - a_1\mathbf{4}^-\mathbf{1}^+ + a_3(s + \mathbf{1}^- - \mathbf{4}^-)\mathbf{3}^+ + a_4\mathbf{1}^-\mathbf{4}^+$$
$$+ a_6(s + \mathbf{1}^- + \mathbf{2}^- - \mathbf{4}^- - \mathbf{9}^-)\mathbf{6}^+ + a_7(s + t + \mathbf{1}^- + \mathbf{2}^- - \mathbf{4}^- - \mathbf{9}^-)\mathbf{7}^+$$
$$- a_8(t - \mathbf{1}^- + \mathbf{4}^-)\mathbf{8}^+ = 0$$

$$5)IBP[k_2,k_2]: -a_1 + a_6 - a_7 + a_8 + a_1\mathbf{8}^-\mathbf{1}^+ - a_3(m_t^2 - s - t + \mathbf{1}^- - \mathbf{8}^-)\mathbf{3}^+ + a_4(t - \mathbf{1}^- + \mathbf{8}^-)\mathbf{4}^+$$
$$- a_6(m_w^2 + \mathbf{7}^-)\mathbf{6}^+ - a_7(m_w^2 - \mathbf{6}^-)\mathbf{7}^+ - a_8\mathbf{1}^-\mathbf{8}^+ = 0$$

$$6)IBP[k_2,p_1]: a_2 - a_5 + a_7 - a_9 - a_2(m_t^2 - m_w^2 - s - t + \mathbf{1}^- + \mathbf{3}^- + \mathbf{5}^- - \mathbf{7}^- - \mathbf{8}^- + \mathbf{9}^-)\mathbf{2}^+$$
$$- a_5(m_t^2 - m_w^2 - s - t - \mathbf{2}^- + \mathbf{4}^- - \mathbf{7}^- + \mathbf{9}^-)\mathbf{5}^+$$
$$- a_6(m_t^2 - m_w^2 - s - t - \mathbf{2}^- + \mathbf{5}^- - \mathbf{7}^- - \mathbf{8}^- + \mathbf{9}^-)\mathbf{6}^+$$
$$- a_7(m_t^2 - m_w^2 - s - t - \mathbf{1}^- - \mathbf{2}^- + \mathbf{5}^- + \mathbf{9}^-)\mathbf{7}^+$$
$$- a_9(m_t^2 - m_w^2 - s - t + 2\,\mathbf{1}^- - \mathbf{2}^- + \mathbf{3}^- - \mathbf{4}^- + \mathbf{5}^- - \mathbf{7}^- - \mathbf{8}^-)\mathbf{9}^+ = 0$$

$$7)IBP[k_2,p_2]: -2a_2 - a_5 - a_6 - a_7 - a_9 + d - a_5\mathbf{2}^-\mathbf{5}^+ - a_6(\mathbf{2}^- - \mathbf{3}^-)\mathbf{6}^+$$
$$+ a_7(m_t^2 - m_w^2 - s - t + \mathbf{1}^- - \mathbf{2}^- + \mathbf{3}^- - \mathbf{8}^-)\mathbf{7}^+ - a_9\mathbf{2}^-\mathbf{9}^+ = 0$$

$$8)IBP[k_2,q_2]: 2a_2 - a_5 - a_6 + a_7 - a_9 - a_2(m_t^2 - m_w^2 - s - t + \mathbf{1}^- + \mathbf{5}^- + \mathbf{6}^- - \mathbf{7}^- - \mathbf{8}^- + \mathbf{9}^-)\mathbf{2}^+$$
$$- a_5(2m_t^2 - m_w^2 - s - 2t + \mathbf{1}^- - 2\,\mathbf{2}^- + \mathbf{6}^- - \mathbf{7}^- - \mathbf{8}^- + \mathbf{9}^-)\mathbf{5}^+$$
$$- a_6(m_t^2 - m_w^2 - s - t - 2\,\mathbf{2}^- + \mathbf{3}^- + \mathbf{5}^- - \mathbf{7}^- - \mathbf{8}^- + \mathbf{9}^-)\mathbf{6}^+$$
$$- a_7(2m_t^2 - m_w^2 - 2s - 2t - 2\,\mathbf{2}^- + \mathbf{3}^- + \mathbf{5}^- + \mathbf{6}^- - \mathbf{8}^- + \mathbf{9}^-)\mathbf{7}^+$$
$$- a_9(m_t^2 - m_w^2 - 2s - t + \mathbf{1}^- - 2\,\mathbf{2}^- + \mathbf{5}^- + \mathbf{6}^- - \mathbf{7}^- - \mathbf{8}^-)\mathbf{9}^+ = 0$$

$$9)\, IBP[k_1, k_2] : -a_2 + a_9 + a_2 9^- 2^+ - a_5(s + t + 2^- - 9^-)5^+$$
$$- a_6(s + 1^- + 2^- - 4^- - 9^-)6^+$$
$$- a_7(s + t + 1^- + 2^- - 4^- - 9^-)7^+ - a_9 2^- 9^+ = 0$$
$$10)\, IBP[k_2, k_1] : -a_6 + a_7 - a_2(m_t^2 - m_w^2 - s - t + 1^- + 6^- - 7^- - 8^-)2^+$$
$$+ a_5(m_w^2 + t - 1^- - 6^- + 7^- + 8^-)5^+$$
$$+ a_6(m_w^2 + 7^-)6^+ + a_7(m_w^2 - 6^-)7^+$$
$$- a_9(m_t^2 - m_w^2 - s + 1^- + 6^- - 7^- - 8^-)9^+ = 0, \tag{5.61}$$

where m_t is the mass of the top quark and m_w is the mass of the W boson. s and t are the Mandelstam variables and d the space-time dimension. There are two kinds of propagators

$$\mathbf{i}^+ I(a_1, \cdots, a_i, \cdots a_9) := I(a_1, \cdots, a_i + 1, \cdots a_9),$$
$$\mathbf{i}^- I(a_1, \cdots, a_i, \cdots a_9) := I(a_1, \cdots, a_i - 1, \cdots a_9), \tag{5.62}$$

as defined in eq.(2.68).

The notation $IBP[v_1, v_2]$, $(v_1, v_2 \in \{k_1, k_2, p_1, p_2, q_2\})$, means: the IBP identity belonging to

$$\int d^d k_1 d^d k_2 \frac{\partial}{\partial v_1^\mu} \left[v_2^\mu \, I'(k_1, k_2, p_1, p_2, q_2) \right] = 0. \tag{5.63}$$

Now we use the integrals in eq.(5.60) as seeds and put them in the IBP relations in eq.(5.61). So we put for example $I(1, 1, 1, 1, 1, 1, 1, -2, -2)$ in the $IBP[k_1, k_1]$ eq.(5.61) and get the following equation

$$IBP[k_1, k_1] : (d - 3)I(1, 1, 1, 1, 1, 1, 1, -2, -2) + 2I(0, 1, 1, 1, 1, 1, 1, -1, -2) - I(0, 1, 1, 1, 1, 1, 2, -2, -2)$$
$$- I(0, 1, 1, 2, 1, 1, 1, -2, -2) - I(0, 1, 2, 1, 1, 1, 1, -2, -2) - I(1, 0, 1, 1, 1, 1, 2, -2, -2)$$
$$- I(1, 0, 1, 1, 1, 2, 1, -2, -2) + I(1, 1, 1, 1, 0, 1, 2, -2, -2) + I(1, 1, 1, 1, 0, 2, 1, -2, -2)$$
$$+ I(1, 1, 1, 1, 1, 1, 2, -2, -3) - (-m_t^2 + m_w^2 + s + t)I(1, 1, 1, 1, 1, 1, 2, -2, -2)$$
$$- I(1, 1, 1, 1, 1, 2, 0, -2, -2) - I(1, 1, 1, 1, 1, 2, 1, -3, -2) + I(1, 1, 1, 1, 1, 2, 1, -2, -3)$$
$$- (-m_t^2 + m_w^2 + s + t)I(1, 1, 1, 1, 1, 2, 1, -2, -2) = 0$$
$$\tag{5.64}$$

In the last step we solve eq.(5.64) for the integral $I(1, 1, 1, 1, 1, 1, 1, -2, -2)$ and consider all other integrals with 6 propagators, one dot and four scalar products on the right hand side as additional seeds for $-\{\mathbf{r} : [\mathbf{t}, \mathbf{t}],\ s : [4, 4]\}$ in the sixth sector. We omit all other integrals with more than one dot, and more than four scalar products . We consider for example the integral $I(0, 1, 1, 1, 1, 1, 2, -2, -2)$ as a additional seed but not the $I(1, 1, 1, 1, 1, 2, 1, -3, -2)$. This step is repeated for all other IBP relations in eq.(5.61). With these seeds we can run the reduction recursively for all sixth sub-sectors again and following the 7_127. We applied this strategy to topologies 1,3,5 and 6 (see fig. 5.3). At the time of submitting this thesis, the reduction of the last sector, 7_127, for 4 scalar products, is still running for two topologies, 1 and 6. However, the final results might be so large, that we would not be able to use them further.

5.3.2. Results

We used the projection method, described in subsec. 2.1.3, to calculate all double boxes. One may not use the Tarasov-method, described in subsec. 2.1.2, for this calculation because through this method we get up to 8 dots and so we can not reduce all occurring scalar integrals to master integrals. Six diagrams belong to the planar double boxes and 12 to non-planar double boxes. We can see in the fourth column of tab. 5.2 how big the final results are, when m_w and m_t are treated as independent variables. They are huge, therefore we can not present them here.

Diagram	Topology	Number of master integrals	Size of diagram	Size of diagram $m_w^2 = \frac{3}{14} m_t^2$
37	$d1$	71	r	25M
47*	$d1$	71	r	26M
40*	$d2$	35	5.4M	1.8M
50*	$d2$	35	6.7M	2.1M
35*	$d3$	77	147M	29M
46	$d3$	77	148M	28M
36*	$d4$	59	63M	22M
45	$d4$	59	62M	21M
38*	$d5$	65	94M	17M
49	$d5$	65	99M	17M
39	$d6$	80	r	57M
48*	$d6$	80	r	57M
53	$d7$	49	19M	4.3M
55*	$d7$	49	19M	4.4M
52*	$d8$	31	2.9M	883K
56*	$d8$	31	2.7M	819K
51	$d9$	51	29M	4.7M
54*	$d9$	51	30M	4.8M

Table 5.2.: The final results for double boxes. $d_1, \cdots, d_9$ are the topology numbers as shown in fig. 5.3 and fig. 5.2. The diagrams which are reduced are marked with an asterisk. r: the reduction is still running in the last sector, 7_127, for 4 scalar products

One reason why the reduction to master integrals is so complicated is that we have five variables s, t, m_t, m_w and d. We can replace the mass of the W boson m_w by a rather exact numerical ratio between the mass of the top quark and the mass of the W boson. For this aim we use the numerical values of the masses from [116]:

$$m_w = 80.385 \pm 0.015 \; GeV$$
$$m_t = 173.5 \pm 0.6 \pm 0.8 \; GeV \tag{5.65}$$

and with the same value for m_w but taking for m_t the value from [117]:

$$m_t = 173.18 \pm 0.94 \; GeV. \tag{5.66}$$

That means we can replace the mass of the W boson as follows

$$m_w^2 \approx 0.214 m_t^2 \approx \frac{3}{14} m_t^2. \tag{5.67}$$

So we have in this case one variable fewer than before. That makes the step of the reduction to master integrals easier. This provides us not only with a consistency check for the reduction tables but we can also calculate all diagrams again with this estimate. Although the final expressions for the fixed $m_w - m_t$ ratio are considerably shorter, they are still too long to be presented here. However, we prepare a C code which can be used as a static library. This code and a guidance which contains details to the C code and the list of all master integrals can be downloaded from:
http://people.physik.hu-berlin.de//~assadsol.

As we can see in tab. 5.2, the size of the results correlates to the number of master integrals. No matter whether we did the reduction for the fixed $m_w - m_t$ ratio or for the whole problem with m_w and m_t, we got the same number of master integrals.

For the application of the IBP method as described in subsec. 2.3.3 we classify the diagrams according to their topology. It is sufficient to apply the IBP reduction to one of the two diagrams belonging to each topology. This diagram is marked with an asterisk in tab. 5.2. Then we find a map between two diagrams. With the help of this map we can use the reduction table of the first diagram for the other one. This map can be applied either to the diagram or to the integral table. As a consistency check we did both and compared the results. Furthermore, in some cases we reduced both diagrams to check our setup for the shift of a diagram to the other one.

In the following subsec. 5.3.2.1 we explain the generation of the C code which can be used to calculate the coefficients of each master integral and spin structure.

5.3.2.1. C code

In order to evaluate our expressions numerically we wrote a C code which can be used as a static library. The results are quite large. This fact motivates us to divide the final result for each diagram in a physically motivated way. The final result of each double box diagram can be factorized in coefficients of the spin structures and the master integrals. In other words, the final expression belonging to a diagram can be considered as a matrix

$$dia = \sum_{i=1}^{11} \sum_{j=1}^{518} S_i \, c_{ij} \, MI_j. \tag{5.68}$$

There are in total 518 master integrals (see tab. 5.2) and 11 spin structures as in eq. (5.4). In the next step we consider the Laurent expansion of each coefficient $c_{i,j}$ for $i \in \{1, \cdots, 11\}$

and $j \in \{1, \cdots, 518\}$ in this matrix from ϵ^{-5} to ϵ^4

$$
\begin{array}{c}
 \quad MI_1 \qquad\qquad \cdots \qquad\qquad MI_{518} \\[4pt]
\begin{array}{c} S_1 \\[6pt] \vdots \\[6pt] S_{11} \end{array}
\left(
\begin{array}{ccc}
c_{1,1,-5}\epsilon^{-5} + \cdots + c_{1,1,4}\epsilon^4 & \cdots & c_{1,518,-5}\epsilon^{-5} + \cdots + c_{1,518,4}\epsilon^4 \\
& \ddots & \\
c_{11,1,-5}\epsilon^{-5} + \cdots + c_{11,1,4}\epsilon^4 & \cdots & c_{11,518,-5}\epsilon^{-5} + \cdots + c_{11,518,4}\epsilon^4
\end{array}
\right) ,
\end{array}
\tag{5.69}
$$

or in the other notation

$$
dia = \sum_{i=1}^{11} \sum_{j=1}^{518} \sum_{k=-5}^{4} S_i \, c_{ijk} \, MI_j \, \epsilon^k .
\tag{5.70}
$$

Every entry can be represented clearly by a 3 dimensional array $c[spin][integral][\epsilon]$. In order to be able to assign every matrix to a diagram we need two more indices. Finally, each entry is represented by a five-dimensional array $c[topology][diagram][spin][integral][\epsilon]$. There are 9 topologies. Two diagrams belong to each topology .

There are two libraries *"dblib1.a"* and *"dblib2.a"* which contain all expressions needed for the calculation of the coefficients c_{ijk} in eq.(5.70). Every coefficient, $c[topology][diagram]$ $[spin][integral][\epsilon]$, depends on the Mandelstam variables s and t and the mass of the top quark. Using these libraries *"dblib1.a"* and *"dblib2.a"* we get a numerical value for each coefficient up to a global factor

$$
C_{global,DBoxes} = \frac{1}{N} C_F \frac{\alpha_e \alpha_s V_{th} V_{ud}^*}{8 \, \sin^2 \theta_W} \left(\frac{\alpha_s}{4\pi} \right) ,
\tag{5.71}
$$

where $\frac{1}{N} C_F$ is the colour factor (see tab. 5.1), α_s is the coupling constant of the strong interaction and α_e is the fine-structure constant. V_{tb}, V_{ud}^* and θ_W are defined as in eq. (3.6). All integral families belonging to the double box diagrams are listed in App. F.

6

Conclusion

The measurements at LHC are getting more precise. For a satisfactory comparison between theory and experiment higher-order perturbative corrections are essential. The quantum chromodynamics (QCD) corrections usually have a dominant contribution to the higher order corrections. These corrections are typically of the order of several ten percent or more, and even next-to-next-to-leading order (NNLO) QCD corrections are needed for several processes. An NNLO calculation has many building blocks as we can see in fig. 1.2. The main intention of this work is the calculation of two-loop QCD corrections to single top quark production at NNLO.

After a brief introduction to the methods like tensor reduction in section 2.1, integration by parts method in section 2.2 and Feynman Graph Polynomial in section 2.3, we calculated the one-loop corrections squared to single top quark production at NNLO in chapter 3. We performed this calculation with two independent different methods for the naive scheme. The final results in this scheme were equal. In chapter 4 we used our setup to calculate the QCD form factors of heavy quarks at NNLO as a first independent check of the calculation in [3]. We found a complete agreement between our results and the results in the literature for the vector coupling [94, 95] as well as for the axial vector coupling [3]. Lastly, we discussed the two-loop QCD corrections to single top quark production at NNLO. We separated the calculation of two-loop QCD corrections in two parts which were factorized in colour. The first part, including the vertex contributions,was considered in section 5.2, and the second one, including the double boxes, was discussed in section 5.3. We computed the vertex contributions presented in app. C in two independent calculations and with two different methods. Both results were equal. In order to be able to evaluate the final result numerically, a C code is made available which can be used as a static library. Using this library one gets a numerical value for every coefficient, c_{ijk} of each spin structure, S_1 and S_3, and each colour coefficient, $\mathcal{C}_1 = C_F^2 N^2$,$\mathcal{C}_2 = C_F C_A N^2$, $\mathcal{C}_3 = I_2(R) C_F N_H N^2$ and $\mathcal{C}_4 = I_2(R) C_F N_L N^2$ expanded in ϵ as a Laurent series [1]

$$vertices = \sum_{i=1}^{2}\sum_{j=1}^{4}\sum_{k=-4}^{0} S_i \, c_{ijk} \, C_j \, \epsilon^k. \tag{6.1}$$

In the calculation of double boxes, the reduction to master integrals turned out rather difficult compared with all other similar calculations due to the number of variables in the problem. There are five variables occurring in the calculation of double boxes: s, t, m_t, m_w and d, where s and t are the Mandelstam variables, m_t and m_w the masses of the top quark and the W boson and d the space-time dimension. In subsec. 5.3.1 we made some proposals to simplify the reduction process.

For the calculation of the double boxes it was not possible to use both described methods for the tensor reduction in subsec. 2.1.2 and in subsec. 2.1.3. The application of the Tarasov-method from subsec. 2.1.2, has the consequence that the powers of the propagators are getting higher. This means that in our case we would have up to 8 *dots*. We name the increasing of the power of each propagator by one a *dot*. In this case it is very difficult to reduce all scalar integrals to master integrals by using common implementations of the *Laporta* algorithm. Thus, we used just the projection method described in subsec. 2.1.3. In this case, we get up to 4 scalar products. However, it is still difficult to reduce all scalar integrals to master integrals (see tab 5.2). Another attempt was to replace the mass of the W boson m_w by a rather exact numerical ratio between the mass of the top quark and the mass of the W boson. For this aim we used the numerical values of the masses from [116, 117]. That means we can replace the mass of the W boson as follows

$$m_w^2 \approx 0.214 m_t^2 \approx \frac{3}{14} m_t^2. \tag{6.2}$$

As we have seen in tab. 5.2, in this case the reduction was much easier and the results are smaller. There is just one colour coefficient for all double box diagrams (see tab. 5.1). The final result of each double box diagram was factorized in coefficients of the spin structures and the master integrals and then expanded in ϵ

$$dia = \frac{1}{N} C_F \sum_{i=1}^{11}\sum_{j=1}^{518}\sum_{k=-5}^{4} S_i \, c_{ijk} \, MI_j \, \epsilon^k, \tag{6.3}$$

where $\frac{1}{N} C_F$ is the colour factor and S_i, $i \in \{1, \cdots, 11\}$, are the eleven spin structures (see eq.(5.4)) and MI_j, $j \in \{1, \cdots, 518\}$, are the master integrals. We prepared a C code for the numerical evaluation of the coefficients c_{ijk}, appearing in eq. (6.3), which can be used as a static library [2].

The calculation of the master integrals MI_j, see eq. (6.3), for the double box topologies is left as a subject for future research. One may use the method of differential equations [36, 118, 119] for the calculation of the master integrals. A numerical approach would be favored because of the complexity of the integrals occurring in our problem

---

[1] The C code as well as the analytical expressions of the final results can be downloaded from http://people.physik.hu-berlin.de/~assadsol

[2] The C code can be downloaded from http://people.physik.hu-berlin.de/~assadsol

[120, 121]. After computing the master integrals we can calculate the contribution of two-loop corrections interfered with Born at NNLO for the single top quark production, as shown in fig. 1.2.

Calculation of one-loop corrections to single top quark production at NNLO

A.1. Master integrals in vertex diagrams

The following integrals occur in the vertex diagrams:

$$
B_0(v_1, 2, 3) = \frac{1}{i\pi^2} \int d^d k \, \frac{1}{(p_1 - k)^2 (q_2 - k)^2},
$$
$$
C_0(v_1, 1, 2, 3) = \frac{1}{i\pi^2} \int d^d k \, \frac{1}{k^2 (p_1 - k)^2 (q_2 - k)^2},
\tag{A.1}
$$

and

$$
A(v_2, 3) = \frac{1}{i\pi^2} \int d^d k \, \frac{1}{((q_1 - k)^2 - m_t^2)},
$$
$$
B_0(v_2, 1, 3) = \frac{1}{i\pi^2} \int d^d k \, \frac{1}{k^2 ((q_1 - k)^2 - m_t^2)},
$$
$$
B_0(v_2, 2, 3) = \frac{1}{i\pi^2} \int d^d k \, \frac{1}{(p_2 - k)^2 ((q_1 - k)^2 - m_t^2)},
$$
$$
C_0(v_2, 1, 2, 3) = \frac{1}{i\pi^2} \int d^d k \, \frac{1}{k^2 (p_2 - k)^2 ((q_1 - k)^2 - m_t^2)},
\tag{A.2}
$$

where k is the loop momentum and the momenta p_i (incoming) and q_i (outgoing) are the external momenta. q_1 is the momentum belonging to the top quark.

A.2. Box diagrams

On the next pages, we present the final result of the box contributions to single top quark production at NNLO except for some factors, while m_t is set equal to 1.

Here t and s are the Mandelstam variables

$$p_1 \cdot p_2 = \frac{s}{2}, \qquad p_2 \cdot q_2 = -\frac{t}{2}, \tag{A.3}$$

and d the space-time dimension.

$$\sum_{\text{box diagrams}} \mathcal{A}_i =$$

$$
\begin{aligned}
& I_{b_3}(d,1,0,0,1)\,((2\,d\,S_1)/(s^2+s\,t)-(2\,((12-4\,d+d^2)\,s+(d^2-4)\,t-2\,(s+t)\,((20-6\,d+d^2)\,s+(d^2-4)\,t)-m_w^2\,((12-4\,d+d^2)\,s+(d^2-4)\\
& \times\,t)+(s+t)\,(m_w^2\,((4+d^2)\,s+(d^2-4)\,t)+(s+t)\,((28-8\,d+d^2)\,s+(d^2-4)\,t)))\,S_2)/((d-4)\,s^2\,(-1+s+t)\,(s+t)\,(-1+m_w^2+s+t))+(2\\
& \times\,(8-2\,d+2\,(d-4)\,m_w^2+(-20+7\,d)\,s+(-22+7\,d)\,t+m_w^2\,(-3\,(d-2)\,s+(14-5\,d)\,t)+(s+t)\,(m_w^2+s+t)\,((-2+3\,d)\,s+3\\
& \times\,(d-2)\,t)-2\,(s+t)\,((-7+4\,d)\,s+2\,(-5+2\,d)\,t))\,S_3)/((d-4)\,s\,(-1+s+t)\,(s+t)\,(-1+m_w^2+s+t))+(4\,(3\,s^3+(d-2)\,(t-1)\\
& \times\,t\,(-1+m_w^2+t)+s^2\,(-5+4\,t+d\,(m_w^2+t))+s\,(2\,m_w^2\,(-1+(d-1)\,t)+(t-1)\,(-2+(-1+2\,d)\,t)))\,S_4)/((d-4)\,s^2\,(-1+s+t)\\
& \times\,(s+t)\,(-1+m_w^2+s+t))-S_5/(s^2+s\,t)+((d-2)\,S_6)/((d-4)\,s^2)+S_7/(4\,s-d\,s))+(I_{b_4}(d,1,0,0,1)\,((-2\,s\,(2\,(d-2)\,s^3+s^2\,(2\,d\\
& \times\,(-1+m_w^2-t)+d^2\,(-1+3\,t)-4\,(-2+m_w^2+4\,t))+t\,(d^2\,(t-1)\,(-1+m_w^2+t)+2\,d\,t\,(-5+m_w^2+5\,t)+4\,(-1+m_w^2+10\,t-3\,m_w^2\\
& \times\,t-9\,t^2))+s\,(4\,(-1+m_w^2+8\,t-12\,t^2)+d\,(-8\,m_w^2\,t+6\,t^2)+d^2\,(1-5\,t+4\,t^2+m_w^2\,(-1+3\,t))))\,S_1)/((s+t)\,(-1+m_w^2+s+t))+(2\\
& \times\,(-((d-2)\,(2\,(t-3)+d\,(t-1))\,(t-1)\,t\,(-1+m_w^2+t))+s^3\,(d^2\,(t-1)+2\,d\,(2+t)-4\,(1+6\,t))+s\,(-(d^2\,(t-1)\,(-1+m_w^2+t^2))+2\\
& \times\,d\,(2-5\,t+2\,t^2+t^3+m_w^2\,(-2+9\,t-3\,t^2))+4\,(-((t-1)^2\,(1+4\,t))+m_w^2\,(1-6\,t+t^2)))+s^2\,(8-4\,m_w^2+44\,t-44\,t^2+d^2\,(t-1)\\
& \times\,(-2+m_w^2+t)+d\,(m_w^2\,(4-6\,t)+4\,(t-2+t^2))))\,S_2)/((-1+s+t)\,(-1+m_w^2+s+t))+(2\,s\,(2\,(2\,s^3\,(t-1)+s^2\,(8-12\,t+t^2)-(t-1)^2\\
& \times\,(-4+3\,t+3\,t^2)-s\,(10-21\,t+7\,t^2+4\,t^3)-m_w^2\,(t-1)\,(-4-2\,s^2+3\,t+3\,t^2+s\,(6+t)))+d\,((-1+s+t)^2\,(-4+2\,s+t+3\,t^2)+m_w^2\\
& \times\,(2\,s^2+(t-1)^2\,(4+3\,t)+s\,(t-6+3\,t^2))))\,S_3)/((-1+s+t)\,(-1+m_w^2+s+t))+(4\,t\,(4-8\,s+5\,s^2-s^3-10\,t+13\,s\,t-4\,s^2\,t+8\\
& \times\,t^2-5\,s\,t^2-2\,t^3+2\,m_w^2\,(-2+2\,s+s^2+3\,t-t^2)+d\,((-2+s+t)\,(-1+s+t)^2+m_w^2\,(2+s\,(t-3)-3\,t+t^2)))\,S_4)/((-1+s+t)\\
& \times\,(-1+m_w^2+s+t))+s\,(d\,(-2+s+3\,t)-2\,(-2+s+4\,t))\,S_5+(d-2)\,(2\,s-s^2+(t-2)\,t)\,S_6-s\,t\,(-2+s+t)\,S_7))/(2\,(d-4)\,s^2\\
& \times\,t^2)+(I_{b_4}(d,0,0,0,1)\,((2\,(d-2)\,s\,(d^2\,(t-1)\,(-1+m_w^2+s+t)-4\,(1-2\,s+s^2-10\,t+7\,s\,t+9\,t^2+m_w^2\,(-1+s+3\,t))+2\,d\,(s^2+s\\
& \times\,(-1+m_w^2+4\,t)+t\,(-5+m_w^2+5\,t)))\,S_1)/(-1+m_w^2+s+t)+(2\,(d-2)\,((d-2)\,(2\,(t-3)+d\,(t-1))\,(t-1)\,t\,(-1+m_w^2+t)+s^3\\
& \times\,(4+40\,t+d^2\,(1+t)-2\,d\,(2+7\,t))+s^2\,(d^2\,(-2-t+3\,t^2+m_w^2\,(1+t))+d\,(8+20\,t-28\,t^2-2\,m_w^2\,(2+3\,t))+4\,(-2-19\,t+19\\
& \times\,t^2+m_w^2\,(1+4\,t)))+s\,(d^2\,(t-1)\,(-1-2\,t+3\,t^2+m_w^2\,(1+2\,t))+4\,((t-1)^2\,(1+8\,t)+m_w^2\,(-1+2\,t+3\,t^2))-2\,d\,(2+t-10\,t^2+7\\
& \times\,t^3+m_w^2\,(-2+3\,t+3\,t^2))))\,S_2)/((-1+s+t)\,(-1+m_w^2+s+t))+(2\,s\,(2\,(2-d)\,(1-s)\,(d\,(-s+(t-1)^2)-2\,(t-1)\,(-1+s+2\\
& \times\,t))+((d-2)\,(2\,(d-2)+2\,(d-2)\,s^2+2\,(d-3)\,s\,t-3\,(d-2)\,t^2+(d-2)\,m_w^2\,(2\,s+t)-(d-2)\,(2\,m_w^2+4\,s+3\,t)-(d-2)\,t^2\\
& \times\,(m_w^2+s+t)\,(5\,s+3\,t)+t\,((d-2)\,m_w^2\,(3\,s+4\,t)-2\,t\,(9\,s+7\,t)+d\,(s^2+10\,s\,t+7\,t^2))))/(-1+m_w^2+s+t))\,S_3)/(-1+s+t)-(4\,t\\
& \times\,(d\,(-3\,s^3+s^2\,(13-10\,t)-4\,(t-2)\,(t-1)^2+s\,(-18+29\,t-11\,t^2)+2\,m_w^2\,(-4+5\,s+s^2+6\,t-s\,t-2\,t^2))-2\,(-s^3+s^2\,(5-4\\
& \times\,t)-2\,(t-2)\,(t-1)^2+s\,(-8+13\,t-5\,t^2)+2\,m_w^2\,(-2+2\,s+s^2+3\,t-t^2))+d^2\,((-2+s+t)\,(-1+s+t)^2+m_w^2\,(2+s\,(t-3)-3\\
& \times\,t+t^2)))\,S_4)/((-1+s+t)\,(-1+m_w^2+s+t))-(d-2)\,s\,(d\,(-2+s+3\,t)-2\,(-2+s+4\,t))\,S_5+(d-2)^2\,(-2\,s+s^2-(t-2)\\
& \times\,t)\,S_6+(d-2)\,s\,t\,(-2+s+t)\,S_7))/(4\,(12-7\,d+d^2)\,m_w^2\,s^2\,t^2)+((1-s-t+m_w^2\,(-1+2\,s+2\,t))\,I_{b_3}(d,1,0,0,0)\,((-2\,(d-2)\,s\\
& \times\,(-1+s+t)\,(-2\,(d-2)\,s^2+(d-4)\,d\,t\,(-1+m_w^2+t)+(d-2)\,s\,(2-2\,m_w^2+(d-8)\,t))\,S_1)/(-1+m_w^2+s+t)+(2\,(d-2)\,t\,((40-14\\
& \times\,d+d^2)\,s^3+(d^2-4)\,(t-1)\,t\,(-1+m_w^2+t)+s^2\,(-64+16\,m_w^2+d\,(24-6\,m_w^2-28\,t)+76\,t+d^2\,(-2+m_w^2+3\,t))+s\,(4\,(6+3\,m_w^2\\
& \times\,(t-2)-14\,t+8\,t^2)+d^2\,(1-4\,t+3\,t^2+m_w^2\,(-1+2\,t))-2\,d\,(5-12\,t+7\,t^2+m_w^2\,(-5+3\,t))))\,S_2)/(-1+m_w^2+s+t)-(2\,s\,(2\\
& \times\,(d-2)\,s^3\,(-2+d+2\,t)+(d-2)\,(t-1)\,t\,(-1+m_w^2+t)\,(8-6\,t+d\,(-2+3\,t))+s^2\,(2\,d\,(8+2\,m_w^2\,(t-2)-5\,t-2\,t^2)-4\,(4+2\,m_w^2\\
& \times\,(t-1)-3\,t+t^2)+d^2\,(-4+2\,m_w^2+2\,t+3\,t^2))+(d-2)\,s\,(-2\,(2+3\,t-9\,t^2+4\,t^3+m_w^2\,(-2+2\,t+t^2))+d\,(2-8\,t^2+6\,t^3+m_w^2\\
& \times\,(-2+2\,t+3\,t^2))))\,S_3)/(-1+m_w^2+s+t)-(4\,t\,(-((10-7\,d+d^2)\,s^3)-(d-2)\,s^2\,(9-2\,m_w^2+d\,(t-2)-8\,t)+(d-2)^2\,(t-1)\,t\\
& \times\,(-1+m_w^2+t)+(d-2)\,s\,(4-4\,m_w^2-5\,t+t^2+d\,(1+t)\,(-1+m_w^2+t)))\,S_4)/(-1+m_w^2+s+t)-(d-2)\,s\,(-1+s+t)\,((d-2)\,s-(d-4)\\
& \times\,t)\,S_5+(d-2)^2\,(-1+s+t)\,(s^2-t^2)\,S_6+(d-2)\,s\,t\,(-1+s+t)\,(s+t)\,S_7))/(4\,(12-7\,d+d^2)\,m_w^2\,(-1+m_w^2)\,s^2\,t^2\,(-1+s+t)^2)+
\end{aligned}
$$

$$
\begin{aligned}
&(I_{b_4}(d,0,1,1,0)((-2s(-1+s+t)(-2d(-1+m_w^2+s-t)-12t+d^2(-1+m_w^2+s+t))S_1)/t-(2((d^2-4)(t-1)^2t(-1+m_w^2+t)+(d-2)\\
&\times s^3(-2+d-12t+3dt)+s(t-1)(-4(-1+m_w^2)(t-1)-2d(-2-5t+7t^2+m_w^2(2+t))+d^2(-1-4t+5t^2+m_w^2(1+2t)))+s^2\\
&\times(d^2(-2-5t+7t^2+m_w^2(1+t))+4(-2-5t+7t^2+m_w^2(1+2t))+d(8+24t-32t^2-2m_w^2(2+3t))))S_2)/((t-1)t)+(2s(d\\
&\times((t-1)^3(2+3t)+s^2(-2-5t+5t^2)+2s(2+t-7t^2+4t^3)+m_w^2(t-1)(-2-t+3t^2+s(2+5t)))-2((t-1)^3(2+3t)+s^2(-2-8\\
&\times t+7t^2)+s(4+5t-19t^2+10t^3)+m_w^2(t-1)(-2-t+3t^2+s(2+5t))))S_3)/((t-1)t)+(4(2-14s+15s^2-3s^3-6t+28s\\
&\times t-15s^2t+6t^2-14st^2-2t^3-2m_w^2(t-1)(-1+2s+t)+d(s^3+5s^2(t-1)+5s(t-1)^2+(t-1)^3+m_w^2(t-1)(-1+2s+t)))\\
&\times S_4)/(t-1)+((d-2)s(-1+s+t)(-1+m_w^2+s+t)S_5)/t+((-1+s+t)((d-2)(t-1)t(-1+m_w^2+t)+s^2(-2+d-12t+3dt)+s\\
&\times(d(-1-3t+4t^2+m_w^2(1+t))-2(-1-6t+7t^2+m_w^2(1+2t))))S_6)/((t-1)t)-s(-1+s+t)(-1+m_w^2+s+t)S_7))/(2(d-4)\\
&\times s^2(-1+s+t)^2)+(I_{b_3}(d,0,1,1,0)((2ds(-1+s+t)(-1+m_w^2+s+t)S_1)/t-(2((d^2-4)(t-1)^2t(-1+m_w^2+t)+(8-6d+d^2)s^3\\
&\times(-1+3t)+s^2(t-1)(d^2(-2+m_w^2+7t)+4(-4+2m_w^2+7t)-2d(-6+3m_w^2+16t))+s(t-1)(-2d(3+m_w^2(t-3)-10t+7t^2)-4\\
&\times(2(t-1)+m_w^2(2+t))+d^2(1-6t+5t^2+m_w^2(-1+2t))))S_2)/((d-4)(t-1)t)+(2s(d(-1+m_w^2+s+t)(2-5t+3t^2+s(-2+5\\
&\times t))-2((t-1)^2(-4+3t)+s^2(-4+7t)+2s(4-9t+5t^2)+m_w^2(4-7t+3t^2+s(-4+5t))))S_3)/((d-4)t)+(4(-2+d-(d-2)\\
&\times m_w^2+11s-4ds+4t-2dt+(d-2)m_w^2(2s+t)+(s+t)(3(d-3)s+(d-2)t))S_4)/(d-4)-(s(-1+s+t)(-1+m_w^2+s+t)\\
&\times S_5)/t+((-1+s+t)((d-2)(t-1)t(-1+m_w^2+t)+(d-4)s^2(-1+3t)+s(t-1)(4-4m_w^2-14t+d(-1+m_w^2+4t)))S_6)/((d-4)\\
&\times(t-1)t)-(s(-1+s+t)(-1+m_w^2+s+t)S_7)/(d-4)))/(2s^2(-1+s+t)^2)+(I_{b_3}(d,1,0,1,0)((-2ds(-1+s+t)(s-t+2m_w^2s\\
&\times(s+t)-(s-t)(m_w^2+s+t))S_1)/(t(s+t))+(2(-((8-6d+d^2)s^4(-1+2m_w^2-t))+(d^2-4)(t-1)^2t^2(-1+m_w^2+t)+s(t-1)\\
&\times t(d^2t(-4+m_w^2+4t)+2d(-1+6t-5t^2+m_w^2(1+t))+4(5-8t+3t^2+m_w^2(-5+2t)))-s^3(2d(-6+m_w^2(9-15t)+t+11\\
&\times t^2)+d^2(2-4t^2+m_w^2(-3+5t))+4(4+t-9t^2+2m_w^2(-3+5t)))+s^2(t-1)(d^2(-1+m_w^2(1-3t)+6t^2)-4(2+6t-11t^2+m_w^2\\
&\times(-2+3t))+d(6-26t^2+m_w^2(-6+20t))))S_2)/((d-4)(t-1)t(s+t))+(2s(s^3(8-16m_w^2+d(-2+4m_w^2-5t)+10t)-(t-1)t\\
&\times(-1+m_w^2+t)(8-6t+d(-2+3t))+s^2(-16+m_w^2(24-38t)-4t+26t^2+d(4+6t-13t^2+m_w^2(-6+7t)))+s(-(d(2+3t-16\\
&\times t^2+11t^3+m_w^2(-2+5t)))+2(4+t-16t^2+11t^3+m_w^2(-4+13t-8t^2))))S_3)/((d-4)t(s+t))+(4((d-6)s-(d-2)t+m_w^2\\
&\times(-((d-6)s)+(d-2)t)+(s+t)(5s+2(d-2)t)-(s+t)(m_w^2(4s+(d-2)t)+(s+t)((d-1)s+(d-2)t)))S_4)/((d-4)(s+t))+(s\\
&\times(-1+s+t)((-1+2m_w^2)s^2+t(-1+m_w^2+t)+s(1+m_w^2(-1+2t)))S_5)/(t(s+t))+((-1+s+t)((d-4)s^2(-1+2m_w^2-t)-(d-2)\\
&\times(t-1)t(-1+m_w^2+t)+s(t-1)(4-4m_w^2+d(-1+m_w^2-2t)+6t))S_6)/((d-4)(t-1)t)+(s(-1+s+t)(-1+m_w^2+s+t)\\
&\times S_7)/(d-4)))/(2(1-m_w^2)s^2(-1+s+t)^2)+(I_{b_1}(d,0,1,1,0)((4(-1+s+t)(-((d-2)(-1+s)(-1+m_w^2+s))-(-16+5d)(-1+s)\\
&\times t-2(d-4)t^2)S_1)/(-1+s)-(8t((1-s)(-4+d-(d-4)(m_w^2+3s)+2s(-m_w^2+(d-4)s))-(d-4)(-1+s)(-2+m_w^2+5s)\\
&\times t+(d-4)(1-3s)t^2)S_2)/((-1+s)s)+4(-2+d+(d-2)s(m_w^2+s)-(d-2)(m_w^2+2s)+(13-4d)t+(2(d-3)m_w^2+(-15+4d)s)\\
&\times t+(-17+4d)t^2+t((s+t)(2s-(d-6)t)+m_w^2(2s-(d-4)t)))S_3-(4t(-((1-s)(-4+d(-m_w^2+(-1+s)^2)-2m_w^2(-2+s)+9\\
&\times s-5s^2))+(1-s)(-((d-4)(-2+m_w^2))+(17-4d)s)t+(d-4)(-1+3s)t^2)S_4)/((-1+s)s)+((-1+s+t)((d-2)(-1+s)\\
&\times(-1+m_w^2+s)+(-10+3d)(-1+s)t+2(d-4)t^2)S_5)/(-1+s)+((-1+s+t)(-((d-2)(-1+s)s(-1+m_w^2+s))-(-1+s)(4-4\\
&\times m_w^2-14s+d(-1+m_w^2+4s))t+(d-4)(1-3s)t^2)S_6)/((-1+s)s)-t(-1+s+t)(-1+m_w^2+s+t)S_7))/(2(d-4)t^2(-1+s+t)^2)-
\end{aligned}
$$

A. Calculation of one-loop corrections to single top quark production at NNLO

$$
\begin{aligned}
&(I_{b_2}(d,0,1,1,0)((4(-1+s+t)(d(1-2s+s^2-7t+7st+6t^2+m_w^2(-1+s+2t))-2(1-2s+s^2-11t+11st+10t^2+m_w^2(-1+s+3\\
&\times t)))S_1)/(-1+s)-(8t(d(2s^3+(t-1)^2+3s(t-1)t+s^2(-3+5t)+m_w^2(-1+s+t+st))-2(4s^3+(t-1)^2+s^2(-7+10t)+m_w^2\\
&\times(-1+s^2+t+2st)+s(2-8t+6t^2)))S_2)/((-1+s)s)-(4(2(s^3(t-1)+m_w^2(-1+s)(t-1)(-1+s+2t)-(t-1)^2(-1+6\\
&\times t)+s^2(3-10t+4t^2)+s(-3+17t-17t^2+3t^3))+d(s^3+(t-1)^2(-1+3t)-s^2(3-5t+t^2)-s(-3+10t-8t^2+t^3)+m_w^2\\
&\times(s^2+(t-1)^2-s(2-2t+t^2))))S_3)/(-1+s)+(4t(-2+d-(d-2)m_w^2-2(d-1)s+2s^2-2(d-2)t+7st-2t(m_w^2+t)+d\\
&\times(s^2+t^2+m_w^2(s+t))+s(-((s+t)(2s+(9-2d)t))+m_w^2(-2s+(d-4)t)))S_4)/((-1+s)s)-((-1+s+t)(d(1-2s+s^2-5t+5\\
&\times st+4t^2+m_w^2(-1+s+2t))-2(1-2s+s^2-8t+8st+7t^2+m_w^2(-1+s+3t)))S_5)/(-1+s)+((-1+s+t)((d-2)s^3+(d-2)\\
&\times t(-1+m_w^2+t)+s^2(d(-2+m_w^2+4t)-2(-2+m_w^2+7t))+s(2(-1+m_w^2(1-2t)+6t-6t^2)+d(1+m_w^2(t-1)-3t+3t^2)))\\
&\times S_6)/((-1+s)s)+t(-1+s+t)(-1+m_w^2+s+t)S_7))/(2(d-4)t^2(-1+s+t)^2)-(I_{b_1}(d,0,0,1,1)((4(-1+s+t)((d-2)(-1+s)\\
&\times s(-1+m_w^2+s)+(-1+s)(-6+d+6m_w^2-dm_w^2-10s+4ds)t+(8-3d+(-16+5d)s-2(d-4)m_w^2s)t^2-2(d-4)(-1+m_w)\\
&\times(1+m_w)t^3)S_1)/((-1+m_w)(1+m_w)(-1+s)(s+t))+(8t((d-4)(s-t)-(d-4)(4s^2+m_w^2(s-t)-2t^2)+m_w^2((d-6)s^2+2\\
&\times(7-2d)st-3(d-4)t^2)+(d-4)(5s^3+6s^2t-t^3)+(s+t)(-((d-4)s(s+t)(2s+t))+m_w^2(2s^2+3(d-4)st+2(d-4)t^2)))\\
&\times S_2)/((-1+m_w)(1+m_w)(-1+s)s(s+t))-(4((d-2)s-(d-2)s(m_w^2+2s)-(d-6)t+((d-6)m_w^2+s-2ds)t-5t^2+m_w^2\\
&\times((d-2)s^2+dst-2(d-5)t^2)+(s+t)((d-2)s^2+(-7+2d)st+(-5+2d)t^2)+t(s+t)((s+t)(2s-(d-4)t)+m_w^2(2s+(d-4)\\
&\times t)))S_3)/((-1+m_w)(1+m_w)(s+t))-(4t((13-3d)s^2+(d-4)(s-t)-(d-4)m_w^2(s-t)+(d-3)st+2(d-4)t^2+m_w^2\\
&\times((d-6)s^2+2(7-2d)st-3(d-4)t^2)+(s+t)((-14+3d)s^2-st-(d-4)t^2)+(s+t)(-(s(s+t)((d-5)s+(d-4)t))+m_w^2\\
&\times(2s^2+3(d-4)st+2(d-4)t^2)))S_4)/((-1+m_w^2)(-1+s)s(s+t))+((-1+s+t)(-((d-2)(-1+s)s(-1+m_w^2+s))+(1-s)\\
&\times(-((d-6)(-1+m_w)(1+m_w))+2(d-2)s)t+(-2+d+(10-3d)s+2(d-4)m_w^2s)t^2+2(d-4)(-1+m_w)(1+m_w)t^3)\\
&\times S_5)/((-1+m_w)(1+m_w)(-1+s)(s+t))+((-1+s+t)((d-2)(-1+s)s(-1+m_w^2+s)+(-1+s)(-4+d+4m_w^2-dm_w^2-6s+2d\\
&\times s)t+(d-4)(1-2m_w^2+s)t^2)S_6)/((-1+m_w)(1+m_w)(-1+s)s)+(t(-1+s+t)(-1+m_w^2+s+t)S_7)/(-1+m_w^2)))/(2(d-4)t^2\\
&\times(-1+s+t)^2)+(I_{b_4}(d,1,0,1,1)((-2s(-1+s+t)(2(d-2)s^4+t^2(-1+m_w^2+t)(-4+d(4-6t)+d^2(t-1)+12t)-(d-2)s^3(d\\
&\times(-1+3t)-2(-2+m_w^2+9t))-st(-2d(t-2)(-5+5m_w^2+4t)+4(6+m_w^2(t-6)-6t+t^2)+d^2(4+2m_w^2(t-2)-5t+t^2))-s^2\\
&\times(4-48t+48t^2+4m_w^2(-1+5t)+d^2(1-8t+5t^2+m_w^2(-1+3t))-2d(2-21t+18t^2+m_w^2(-2+9t))))S_1)/(s+t)+(2(-((d-2)\\
&\times(t-1)t^3(-1+m_w^2+t)(d(t-1)+2(1+t)))+s^5(4-d^2(t-1)-40t+2d(-2+7t))+st^2(-(d^2(t-1)(3-8t+5t^2+m_w^2(-3+4\\
&\times t)))-4((t-1)^2(-5+6t)+m_w^2(5-10t+t^2))+2d(-8+25t-24t^2+7t^3+m_w^2(8-15t+3t^2)))+s^4(-(d^2(t-1)(-2+m_w^2+5\\
&\times t))+4(-2+m_w^2(1-4t)+23t-39t^2)+d(m_w^2(-4+6t)+8(1-6t+7t^2)))+s^2t(-(d^2(t-1)(3-12t+10t^2+m_w^2(-3+6t)))+2\\
&\times d(-8+42t-62t^2+28t^3+m_w^2(8-26t+9t^2))-4(-5+31t-58t^2+34t^3+m_w^2(5-21t+9t^2)))+s^3(-(d^2(t-1)(1-8t+10\\
&\times t^2+m_w^2(-1+4t)))+2d(-2+25t-62t^2+42t^3+m_w^2(2-15t+9t^2))-4(-1+18t-62t^2+56t^3+m_w^2(1-12t+11t^2))))\\
&\times S_2)/(s+t)+2s((10+3d(t-1)-6t)(t-1)t^2(-1+m_w^2+t)+2s^4(-2+d+2t)+s^3(8+4m_w^2(t-1)-12t+6t^2+d(-4+2m_w^2+5\\
&\times t+3t^2))+s^2(2(-2+3t+5t^2-3t^3+m_w^2(2-5t+t^2))+d(2-6t-5t^2+9t^3+m_w^2(-2+5t+3t^2)))+st(-2(t-1)(1-13t+7\\
&\times t^2+m_w^2(-1+4t))+d((t-1)^2(1+9t)+m_w^2(-1-3t+6t^2))))S_3+4t(s+t)(-((d-5)s^3)+(d-2)(t-1)t(-1+m_w^2+t)+s^2\\
&\times(-9+2m_w^2-d(t-2)+8t)+s(4-4m_w^2-5t+t^2+d(1+t)(-1+m_w^2+t)))S_4+s(-1+s+t)(s+t)(-1+m_w^2+s+t)((d-2)\\
&\times s-(d-4)t)S_5-(d-2)(-1+s+t)(s+t)(-1+m_w^2+s+t)(s^2-t^2)S_6-st(-1+s+t)(s+t)^2(-1+m_w^2+s+t)S_7))/(4(d-3)s^2\\
&\times t^2(-1+s+t)^2)+(I_{b_4}(d,0,0,1,0)((-2s(-1+s+t)((d-2)(12t(1+t)(s+t)-2d(s-s^2+t+9st+2t^2-m_w^2(s+t)-(m_w^2+9\\
&\times s-t)t(s+t))+d^2(s-s(s-3t)+t-m_w^2(s+t)-t(s+t)(m_w^2+5s+t)))+(2(-((d-2)^2s^3)+(28-20d+3d^2)s^2t+(d-2)^2\\
&\times(s+t)+2(6-5d+d^2)st(m_w^2+2t)+2(6-5d+d^2)t^2(m_w^2+2t)-(d-2)((d-2)s^2+2(-16+5d)st+(-14+5d)t^2+(d-2)\\
&\times m_w^2(s+t))+(d-2)s(s+t)((d-2)s^2+(-14+5d)st+4(-7+2d)t^2+m_w^2((d-2)s+2(d-3)t))))/(1-s))S_1)/(s+t)+
\end{aligned}
$$

$$
\begin{aligned}
&(2 ((4 (d-2) t ((d-6) s^2 +(10-3 d) s^3 +3 (d-6) s t+2 (35-9 d) s^2 t -2 (d-2) t^2 +2 (21-5 d) s t^2 +(d-2) t^3 +(d-2) (s+t)-(d-2) \\
&\times m_w^2 (s+t)+m_w^2 (s+t) (2 s+(d-2) t)+s (s+t) (-((d-6) s^2)+2 (-18+5 d) s t+2 (-11+3 d) t^2 +m_w^2 ((d-2) s+2 (d-3) t))+s^2 \\
&\times (s+t) (m_w^2 (-2 s+(d-4) t)+(d-4) (2 s^2 +3 s t+t^2))))/(1-s)+((2-d) ((d-2) (s+t) ((d-2) s-(2+d) t)-(d-2) m_w^2 (s+t) \\
&\times ((d-2) s-(2+d) t)+(d-2) m_w^2 (s+t) ((d-2) s^2 -2 (1+d) s t-(2+d) t^2)+s ((d-2)^2 s^3 -2 (44-30 d+5 d^2) s^2 t+(-112+110 \\
&\times d-23 d^2) s t^2 +2 (22+15 d-6 d^2) t^3)-2 ((d-2)^2 s^3 +(-18+11 d-2 d^2) s^2 t+(14+7 d-4 d^2) s t^2 -(d^2 -4) t^3)+t^2 (s+t) \\
&\times ((d-2) m_w^2 ((d-4) s^2 +2 (1+d) s t+(2+d) t^2)+(s+t) ((8-6 d+d^2) s^2 +2 (10-5 d+d^2) s t+(d^2 -4) t^2))+t (s+t) ((d-2) m_w^2 \\
&\times (2 (d-3) s^2 +(d-2) s t-(2+d) t^2)+2 ((22-17 d+3 d^2) s^3 +(36-28 d+5 d^2) s^2 t+(-22-d+d^2) s t^2 -(d^2 -4) t^3))))/(1-t)) \\
&\times S_2)/(s+t)-(2 (d-2) s (2 (2 s^4 (t-1)^2 +(t-1)^3 (4+t+3 t^2)+s^3 (-2+32 t-25 t^2 +t^3)-s^2 (6+41 t-69 t^2 +24 t^3 +4 t^4)+s (10+2 \\
&\times t-34 t^2 +25 t^3 -3 t^5)+m_w^2 (t-1) (-4+2 s^3 (t-1)+3 t-2 t^2 +3 t^3 -s^2 t (13+t)+s (6+8 t+3 t^2 -3 t^3)))+d (2 s^4 (t-1)-(t-1)^3 \\
&\times (4+3 t+3 t^2)+s^3 (2-21 t+12 t^2 +3 t^3)+s^2 (6+27 t-44 t^2 +9 t^3 +6 t^4)+s (-10+t+26 t^2 -16 t^3 -4 t^4 +3 t^5)+m_w^2 (t-1) (4+2 \\
&\times s^3 -t-3 t^3 +s^2 t (11+3 t)+s (-6-10 t-3 t^2 +3 t^3)))) S_3)/((-1+s) (t-1))-(4 (d-2) t (-4+8 s-9 s^2 +10 s^3 -5 s^4 +10 t+9 \\
&\times s t-18 s^2 t-s^4 t-6 t^2 -44 s t^2 +32 s^2 t^2 -4 s^3 t^2 -2 t^3 +29 s t^3 -5 s^2 t^3 +2 t^4 -2 s t^4 +2 m_w^2 (t-1) (-2-s^2 +s^3 +t+t^2 +s \\
&\times (2+5 t-t^2))+d (s^4 (1+t)-(t-1)^3 (2+t)+s (t-1)^2 (-5-8 t+t^2)+s^3 (-3-2 t+3 t^2)+s^2 (5+4 t-12 t^2 +3 t^3)+m_w^2 (t-1) \\
&\times (2-t-t^2 +s^2 (1+t)+s (-3-4 t+t^2)))) S_4)/((-1+s) (t-1))-((d-2) s (-1+s+t) (-2 (s^3 +2 (t-1)^2 +13 s^2 t+m_w^2 (-2+s+s^2 +2 \\
&\times t+4 s t)+3 s (-1-3 t+4 t^2))+d (2+s^3 -3 t+8 s^2 t+t^2 +m_w^2 (-2+s+s^2 +t+3 s t)+s (-3-5 t+7 t^2))) S_5)/(-1+s)+((d-2) \\
&\times (s-t) (-1+s+t) (-2 (s^3 (t-1)+2 s^2 (t-8) t-(t-1)^2 (2+t)+s (3+12 t-16 t^2 +t^3)+m_w^2 (2+s^2 (t-1)-t-t^2 +s (-1-6 \\
&\times t+t^2)))+d (s^3 (t-1)+2 s^2 (t-5) t-(t-1)^2 (2+t)+s (3+6 t-10 t^2 +t^3)+m_w^2 (2+s^2 (t-1)-t-t^2 +s (-1-4 t+t^2)))) \\
&\times S_6)/((-1+s) (t-1))+(d-2) s t (-1+s+t) (2+s+t) (-1+m_w^2 +s+t) S_7))/(8 (12-7 d+d^2) s^2 t^2 (-1+s+t)^2)+(I_{b_1}(d,1,1,1,1) \\
&\times (4 s (-1+s+t) (-1+m_w^2 +s+t) (d (-1+m_w^2 +s+3 t)-2 (-1+m_w^2 +s+4 t)) S_1 +8 (1-m_w^2 -s-t) t (-4+d+(d-4) (s+t) \\
&\times (2 s+t)-(d-4) (m_w^2 +3 s+2 t)+m_w^2 (-2 s+(d-4) t)) S_2 +4 s (-((d-2) (-1+s) (-1+m_w^2 +s)^2)-(1-m_w^2 -s) (-7+3 \\
&\times d-2 s (m_w^2 +s)-(d-3) (2 m_w^2 +3 s)) t+(4 (d-3)+(11-3 d) m_w^2 +(d-4) m_w^4 -5 (d-4) s-6 m_w^2 s+(d-8) s^2) t^2 +(11-3 \\
&\times d-2 m_w^4 +2 (d-5) s) t^3 +(d-4) t^4) S_3 -4 (1-m_w^4 -s-t) t (-4+9 s-5 s^2 +8 t-9 s t-4 t^2 -2 m_w^2 (-2+s+2 t)+d (m_w^4 \\
&\times (t-1)+(-1+s+t)^2)) S_4 -(d-2) s (-1+s+t) (-1+m_w^2 +s+t)^2 S_5 +(-1+s+t) (-1+m_w^2 +s+t)^2 ((d-2) s+(d-4) t) S_6 +s t \\
&\times (-1+s+t) (-1+m_w^2 +s+t)^2 S_7))/(4 (d-3) t^2 (-1+s+t)^2)+(I_{b_3}(d,1,1,1,1) (-2 (d-4) d s (-1+s+t) (-1+m_w^2 +s+t)^2 S_1 +2 \\
&\times (-1+m_w^2 +s+t) ((8-6 d+d^2) s^3 +(d^2 -4) (t-1) t (-1+m_w^2 +t)+s^2 (d^2 (-2+m_w^2 +3 t)+4 (-4+2 m_w^2 +7 t)-2 d (-6+3 m_w^2 +8 \\
&\times t))+s (-2 d (3+m_w^2 (t-3)-8 t+5 t^2)-4 (-2+6 t-4 t^2 +m_w^2 (2+t))+d^2 (1-4 t+3 t^2 +m_w^2 (-1+2 t)))) S_2 +2 s (-(d (2 m_w^2 \\
&\times (-1+s+t)^2 (-2+3 t)+m_w^4 (2-5 t+3 t^2 +s (-2+5 t))+(-1+s+t)^2 (2-5 t+3 t^2 +s (-2+5 t))))+2 (m_w^4 (4-7 t+3 t^2 +s (-4+5 \\
&\times t))+(-1+s+t)^2 (4-7 t+3 t^2 +s (-4+5 t))+2 m_w^2 (2 s^2 (t-2)+(t-1)^2 (-4+3 t)+s (8-13 t+5 t^2)))) S_3 +4 (1-m_w^2 -s-t) t \\
&\times (-2+3 s-s^2 +4 t-3 s t-2 t^2 -2 m_w^2 (-1+2 s+t)+d ((-1+s+t)^2 +m_w^2 (-1+2 s+t))) S_4 +(d-4) s (-1+s+t) (-1+m_w^2 +s+t)^2 \\
&\times S_5 -(-1+s+t) (-1+m_w^2 +s+t)^2 ((d-4) s+(d-2) t) S_6 +s t (-1+s+t) (-1+m_w^2 +s+t)^2 S_7))/(4 (d-3) s^2 (-1+s+t)^2)+((1-s) \\
&\times I_{b_2}(d,1,1,1,1) (4 s (1-m_w^2 -s-t) (-1+s+t) (-2+d-(d-2) m_w^2 +4 s-2 d s+14 t-5 d t+m_w^2 ((d-2) s+2 (d-3) t)+(s+t) \\
&\times ((d-2) s+4 (d-3) t)) S_1 -8 (1-m_w^2 -s-t) t (-2+d-(d-2) m_w^2 -4 s-2 (d-2) t-2 (-7 s^2 -4 s t+t (m_w^2 +t))+d (-3 \\
&\times s^2 -s t+t^2 +m_w^2 (s+t))+s (m_w^2 (-2 s+(d-4) t)+(d-4) (2 s^2 +3 s t+t^2))) S_2 +4 s (d (2 m_w^2 (-1+s+t)^3 +m_w^4 (s^2 +(t-1)^2 -s \\
&\times (2-2 t+t^2))+(-1+s+t)^2 (s^2 +(t-1)^2 -s (2-2 t+t^2)))+2 (m_w^4 (-1+s) (t-1) (-1+s+2 t)+(-1+s) (t-1) (-1+s+t)^2 \\
&\times (-1+s+2 t)+m_w^2 (2 s^3 (t-1)-2 (t-1)^2 (-1+2 t)+3 s^2 (2-4 t+t^2)+s (-6+18 t-13 t^2 +t^3)))) S_3 +4 (1-m_w^2 -s-t) t \\
&\times (-2+2 s+2 s^2 -2 s^3 +4 t-s t-3 s^2 t-2 t^2 -s t^2 -2 m_w^2 (-1+s^2 +t+2 s t)+d ((-1+s+t)^2 +m_w^2 (-1+s+t+s t))) S_4 +
\end{aligned}
$$

$$s(-1+s+t)(-1+m_w^2+s+t)^2(d(-1+s+2t)-2(-1+s+3t))S_5+(-1+s+t)(-1+m_w^2+s+t)^2((d-2)(s-t)-s((d-2)s+(d-4)t))$$
$$\times S_6-(-1+s)st(-1+s+t)(-1+m_w^2+s+t)^2 S_7)/(4(d-3)s^2t^2(-1+s+t)^2)+((1-t)I_{b_4}(d,1,1,1,1)(-2s(t-1)(-1+s+t)$$
$$\times(-1+m_w^2+s+t)(-2d(-1+m_w^2+s-t)-12t+d^2(-1+m_w^2+s+t))S_1-2(-1+m_w^2+s+t)((d^2-4)(t-1)^2t(-1+m_w^2+t)+(d-2)$$
$$\times s^3(-2+d-4t+dt)+s(t-1)(-4(-1+m_w^2-4t)(t-1)-2d(-2-3t+5t^2+m_w^2(2+t))+d^2(-1-2t+3t^2+m_w^2(1+2t)))+s^2$$
$$\times(d^2(-2-t+3t^2+m_w^2(1+t))+4(-2-5t+7t^2+m_w^2(1+2t))-2d(-4-4t+8t^2+m_w^2(2+3t))))S_2+2s(d(m_w^4(t-1)(-2-t+3$$
$$\times t^2+s(2+5t))+(-1+s+t)^2((t-1)^2(2+3t)+s(-2-t+5t^2))+2m_w^2((t-1)^3(2+3t)+s^2(-2-2t+3t^2)+s(4-t-9t^2+6$$
$$\times t^3)))-2(m_w^4(t-1)(-2-t+3t^2+s(2+5t))+(-1+s+t)^2((t-1)^2(2+3t)+s(-2+5t^2))+m_w^2(2(t-1)^3(2+3t)+s^2(-4-3$$
$$\times t+4t^2)+s(8-3t-15t^2+10t^3))))S_3+4(1-m_w^2-s-t)t(-2+6s-7s^2+3s^3+6t-12st+7s^2t-6t^2+6st^2+2t^3+2m_w^2$$
$$\times(t-1)(-1+2s+t)-d((-1+s+t)^3+m_w^2(t-1)(-1+2s+t)))S_4+(d-2)s(t-1)(-1+s+t)(-1+m_w^2+s+t)^2 S_5+(-1+s+t)$$
$$\times(-1+m_w^2+s+t)^2((d-2)(t-1)t+s(-2+d-4t+dt))S_6-s(t-1)t(-1+s+t)(-1+m_w^2+s+t)^2 S_7))/(4(d-3)s^2t^2(-1+s+t)^2)$$
$$+I_{b_1}(d,1,0,0,1)((-4((d-2)s^2+(d-2)s(-1+m_w^2+3t)+2t(-1+m_w^2+(d-2)t))S_1)/((d-4)t^2(s+t)(-1+m_w^2+s+t))+(8$$
$$\times(-4+d-(d-4)(m_w^2+3(s+t))+2(s+t)(-m_w^2+(d-4)(s+t)))S_2)/((d-4)t(-1+s+t)(s+t)(-1+m_w^2+s+t))+(4(2(t-1)$$
$$\times t(-1+m_w^2+t)+s^3(-2+d+2t)+s^2(4+2m_w^2(t-1)-5t+4t^2+d(-2+m_w^2+2t))+s((t-1)(2+2m_w^2(t-1)+t+2t^2)+d$$
$$\times((t-1)^2+m_w^2(-1+2t))))S_3)/((d-4)t^2(-1+s+t)(s+t)(-1+m_w^2+s+t))-(4(-4+d-(d-4)m_w^2-(-9+2d)(s+t)+(s+t)$$
$$\times(-2m_w^2+(d-5)(s+t)))S_4)/((d-4)t(-1+s+t)(s+t)(-1+m_w^2+s+t))+(((d-2)s+2t)S_5)/((d-4)t^2(s+t))-((d-2)$$
$$\times S_6)/((d-4)t^2)+S_7/(4t-dt))+(I_{b_1}(d,0,1,1,1)((4((d-2)(-1+s)^2(-1+m_w^2+s)+(1-s)(-8+3d+2(d-4)m_w^2+(8-3d)s)s$$
$$\times t-4(d-3)(1+m_w^2-s)t^2)S_1)/((-1+s)t^2(-1+s+t))-(8((-1+s)^2(-4+d-(d-4)(m_w^2+3s)+2s(-m_w^2+(d-4)s))-(1-s)$$
$$\times(2(d-4)-(d-4)(3m_w^2+5s)+s((4-3d)m_w^2+3(d-4)s))t+(-4+d-2(d-4)(m_w^2+s)+s(-2(d-2)m_w^2+(d-4)s))t^2)$$
$$\times S_2)/((-1+s)st(-1+s+t)^2)+(4(-((d-2)(-1+s)^2(-1+m_w^2+s))-(-1+s)(5+2d(-1+s)-7s+2s^2+2m_w^2(1+s))t+(4$$
$$\times m_w^2(-2+s)+7(-1+s)-d(-1+s)(2+3m_w^2+s))t^2-(-4+d+(d-2)(2m_w^2+s))t^3)S_3)/(t^2(-1+s+t)^2)-(4(-((-1+s)^2$$
$$\times(-4+d(-m_w^2+(-1+s)^2)-2m_w^2(-2+s)+9s-5s^2))+(1-s)(-8+11s-3s^2+4m_w^2(3+s)-d(2(-1+s)+3m_w^2(1+s)))$$
$$\times t+(-1+2m_w^2+s)(-4+d-2s+ds)t^2)S_4)/((-1+s)st(-1+s+t)^2)+((-((d-2)(-1+s)^2(-1+m_w^2+s))-(1-s)(-2+d+2$$
$$\times(d-4)m_w^2-(d-2)s)t+4(d-3)m_w^2t^2)S_5)/((-1+s)t^2(-1+s+t))+(((d-2)(-1+s)^2s(-1+m_w^2+s)-(1-s)(-((d-4)$$
$$\times(-1+m_w)(1+m_w))+(10+4m_w^2-d(3+m_w^2))s+2(d-3)s^2)t+(-4+d-2(d-4)(m_w^2+s)+s(-2(d-2)m_w^2+(d-4)s))t^2)$$
$$\times S_6)/((-1+s)st^2(-1+s+t))+((-1+m_w^2+s)/t+m_w^2/(-1+s+t))S_7))/(4(d-3))+(I_{b_4}(d,0,1,1,1)((2s(-1+s+t)(-12t$$
$$\times(-1+2s+t)+2d(-1+s+3st+t^2-m_w^2(-1+2s+t))+d^2((t-1)(-1+s+t)+m_w^2(-1+2s+t)))S_1)/t+(2((d^2-4)(t-1)^2$$
$$\times t(-1+m_w^2+t)-(d-2)s^3(2-4m_w^2+d(-1+2m_w^2-t)+4t)+s(t-1)(2d(2+m_w^2(t-2)+3t-5t^2)+d^2(-1+m_w^2-2$$
$$\times t+3t^2)+4(-1-3t+4t^2+m_w^2(1+3t)))+s^2(-(d^2(-2+3m_w^2-3t)(t-1))+4(-2+3m_w^2-5t+7t^2)+2d(4+4t-8$$
$$\times t^2+m_w^2(-6+7t)))S_2)/t-(2s(2(-((t-1)^3(2+3t))+s^2(2-6t+t^2)-s(4-9t+3t^2+2t^3)+m_w^2(2s^2(t-2)-(t-1)^2$$
$$\times(2+3t)+s(6-7t+t^2)))+d((t-1)^3(2+3t)+s^2(-2+3t+t^2)+s(4-6t-2t^2+4t^3)+m_w^2(2s^2(2+t)+(t-1)^2(2+3t)+3$$
$$\times s(t-2+t^2))))S_3)/t+4(-2+d+(-13+5d)s^2+6(d-2)st+3(d-2)t^2+2m_w^2(2s+(d-2)t)-m_w^2(4s^2+4st+(d-2)$$
$$\times t^2)-(s+t)(3(d-3)s^2+2(d-2)st+(d-2)t^2)-(d-2)(m_w^2+3(s+t)))S_4-((d-2)s(-1+s+t)((t-1)(-1+s+t)+m_w^2$$
$$\times(-1+2s+t))S_5)/t-((-1+s+t)((d-2)(t-1)t(-1+m_w^2+t)-s(-2+m_w^2(2-4t)+d(-1+m_w^2-2t)(t-1)-4t+6t^2)+s^2$$
$$\times(-2+4m_w^2-4t+d(1-2m_w^2+t)))S_6)/t+s(-1+s+t)((t-1)(-1+s+t)+m_w^2(-1+2s+t))S_7))/(4(d-3)s^2(-1+s+t)^2)+$$

$$
\begin{aligned}
&(I_{b_3}(d,1,1,1,0)\,((-2\,(d-4)\,d\,s\,(-1+s+t)\,((t-1)\,(-1+s+t)+m_w^2\,(-1+2\,s+t))\,S_1)/t+(2\,((d^2-4)\,(t-1)^3\,t\,(-1+m_w^2+t)+\\
&s\,(t-1)^2\,(-(d^2\,(-1+m_w^2+4\,t-3\,t^2))+2\,d\,(-3+8\,t-5\,t^2+m_w^2\,(3+t))+4\,(2-6\,t+4\,t^2+m_w^2\,(-2+3\,t)))+\\
&s^3\,(d^2\,((t-1)^2-2\,m_w^2\,(1+t))-8\,(-(t-1)^2+m_w^2\,(2+t))+2\,d\,(-3\,(t-1)^2+m_w^2\,(6+4\,t)))+s^2\,(t-1)\,(4\,(4-6\,m_w^2-11\,t+7\,t^2)+\\
&d^2\,(2-5\,t+3\,t^2-3\,m_w^2\,(1+t))+2\,d\,(-6+14\,t-8\,t^2+m_w^2\,(9+7\,t))))\,S_2)/((t-1)\,t)-(2\,s\,(d\,((t-1)\,(s^2\,(t-2)+4\,s\,(t-1)^2+\\
&(t-1)^2\,(-2+3\,t))+m_w^2\,(2\,s^2\,(t-2)+(t-1)^2\,(-2+3\,t)+3\,s\,(2-3\,t+t^2)))+2\,((t-1)\,(s^2\,(4+t)-(t-1)^2\,(-4+3\,t)-\\
&2\,s\,(4-5\,t+t^2))+m_w^2\,(2\,s^2\,(4+t)-(t-1)^2\,(-4+3\,t)+s\,(-12+11\,t+t^2))))\,S_3)/t+4\,(-2+d-(d-2)\,m_w^2+(3-2\,d)\,s+(-7+3\,d)\\
&\times\,s^2-3\,(d-2)\,t+2\,(-3+2\,d)\,s\,t+3\,(d-2)\,t^2+2\,m_w^2\,(2\,s+(d-2)\,t)-m_w^2\,(4\,s^2+4\,s\,t+(d-2)\,t^2)-(s+t)\,(2\,(d-3)\,s^2+(d-1)\,s\\
&\times\,t+(d-2)\,t^2))\,S_4+((d-4)\,s\,(-1+s+t)\,((t-1)\,(-1+s+t)+m_w^2\,(-1+2\,s+t))\,S_5)/t-((-1+s+t)\,((d-2)\,(t-1)^2\,t\,(-1+m_w^2\\
&+t)+s^2\,(-4\,(t-1)^2+4\,m_w^2\,(2+t)+d\,((t-1)^2-2\,m_w^2\,(1+t)))-s\,(t-1)\,(4-10\,t+6\,t^2-4\,m_w^2\,(1+t)+d\,(-1+3\,t-2\,t^2+m_w^2\,(1\\
&+t))))\,S_6)/((t-1)\,t)+s\,(-1+s+t)\,((t-1)\,(-1+s+t)+m_w^2\,(-1+2\,s+t))\,S_7))/(4\,(d-3)\,s^2\,(-1+s+t)^2)+(I_{b_2}(d,1,1,1,0)\,(4\,(-1+\\
&s+t)\,(-2+d-(d-2)\,m_w^2+4\,s-2\,d\,s+14\,t-5\,d\,t+m_w^2\,((d-2)\,s+2\,t)+(s+t)\,((d-2)\,s+4\,(d-3)\,t))\,S_1-(8\,t\,(d\,(2\,s^3+3\,s^2\,(t-1)+\\
&(t-1)^2+s\,(t-1)\,t+m_w^2\,(-1+s+3\,t-3\,s\,t-2\,t^2))-2\,(4\,s^3+(t-1)^2+2\,s\,(t-1)^2+s^2\,(-7+6\,t)+m_w^2\,(-1+s^2+3\,t-2\,s\,t-2\,t^2)))\\
&\times\,S_2)/s-4\,(2\,(s^2\,(3-6\,t)+s^3\,(t-1)-(t-1)^2\,(-1+2\,t)-s\,(3-9\,t+5\,t^2+t^3)+m_w^2\,(t-1)\,(1+s^2+2\,t-2\,t^2-2\,s\,(1+t)))\\
&+d\,(s^3+(t-1)^3+s^2\,(-3+3\,t+t^2)+s\,(3-6\,t+2\,t^2+t^3)+m_w^2\,(s^2+(t-1)^2\,(1+2\,t)+s\,(-2+3\,t^2))))\,S_3+(4\,t\,(-2+d-(d-2)\,m_w^2\\
&-2\,(d-1)\,s+2\,s^2-2\,(d-2)\,t-7\,s\,t-2\,t\,(3\,m_w^2+t)-s\,(s+t)\,(2\,s+(-5+2\,d)\,t)+m_w^2\,(-2\,s^2+(4-3\,d)\,s\,t-2\,(d-2)\,t^2)\\
&+d\,(s^2+4\,s\,t+t^2+m_w^2\,(s+3\,t)))\,S_4)/s-(-1+s+t)\,(-2+d-(d-2)\,m_w^2+4\,s-2\,d\,s+8\,t-3\,d\,t+m_w^2\,((d-2)\,s+2\,t)+(s+t)\\
&\times\,((d-2)\,s+2\,(d-3)\,t))\,S_5+((-1+s+t)\,(-2\,(d-2)\,s^2+(d-2)\,(s-t)-(d-2)\,m_w^2\,(s-t)-(d-4)\,s\,t+(d-2)\,t^2+s\,(s+t)\\
&\times\,((d-2)\,s+(d-4)\,t)+m_w^2\,((d-2)\,s^2-(d-4)\,s\,t-2\,(d-2)\,t^2))\,S_6)/s+t\,(-1+s+t)\,((-1+s)\,(-1+s+t)+m_w^2\,(-1+s+2\,t))\\
&\times\,S_7))/(4\,(d-3)\,t^2\,(-1+s+t)^2)
\end{aligned}
$$

$$\tag{A.4}$$

b_i in I_{b_i}, $i \in \{1,2,3,4\}$ labels the box number which one can see in fig.(3.3). We state now the scalar integrals belonging to each diagram:

The scalar integrals belonging to the diagram b_1 and b_3 are:

$$
I_{b_1}(d,a_1,a_2,a_3,a_4)=\int\frac{d^d k}{(2\pi)^d}\,\frac{1}{(k^2)^{a_1}((p_1-k)^2)^{a_2}((p_2+k)^2)^{a_3}((p_1-q_2-k)^2-m_w^2)^{a_4}}\,, \tag{A.5}
$$

and

$$
I_{b_3}(d,a_1,a_2,a_3,a_4)=\int\frac{d^d k}{(2\pi)^d}\,\frac{1}{(k^2-m_w^2)^{a_1}((p_1-k)^2)^{a_2}((q_1-k)^2)^{a_3}((p_1-q_2-k)^2)^{a_4}}\,. \tag{A.6}
$$

As we can see, there is a symmetry between both diagrams b_1 and b_3. This symmetry leads to the following relation between master integrals:

$$
I_{b_1}(d,0,0,0,1)\;\rightarrow\;I_{b_3}(d,1,0,0,0). \tag{A.7}
$$

The scalar integrals belonging to the diagrams b_2 and b_4 are:

$$I_{b_2}(d, a_1, a_2, a_3, a_4) = \int \frac{d^d k}{(2\pi)^d} \frac{1}{(k^2 - m_w^2)^{a_1}((p_1 - k)^2)^{a_2}((p_2 + k)^2 - m_t^2)^{a_3}((p_1 - q_2 - k)^2)^{a_4}},$$
$$\text{(A.8)}$$

and

$$I_{b_4}(d, a_1, a_2, a_3, a_4) = \int \frac{d^d k}{(2\pi)^d} \frac{1}{(k^2)^{a_1}((p_1 - k)^2)^{a_2}((q_1 - k)^2 - m_t^2)^{a_3}((p_1 - q_2 - k)^2 - m_w^2)^{a_4}}.$$
$$\text{(A.9)}$$

There is a symmetry between both diagrams b_2 and b_4 too, which leads to the following relations between master integrals:

$$\begin{aligned}
I_{b_2}(d, 1, 0, 0, 0) &\rightarrow I_{b_4}(d, 0, 0, 0, 1), \\
I_{b_2}(d, 0, 0, 1, 0) &\rightarrow I_{b_4}(d, 0, 0, 1, 0), \\
I_{b_2}(d, 1, 0, 0, 1) &\rightarrow I_{b_4}(d, 1, 0, 0, 1), \\
I_{b_2}(d, 1, 0, 1, 1) &\rightarrow I_{b_4}(d, 1, 0, 1, 1).
\end{aligned}$$
$$\text{(A.10)}$$

We used the relations in eq.(A.7) and eq.(A.10) in the sum of all box diagrams in eq.(A.4).

Final formulas for the dimension shift of the topologies $Sd1$ and $Sd2$

The scalar integral belonging to topology $Sd1$ is:

$$I_{Sd1}(d,a_1,a_2,a_3,a_4,a_5,a_6) = \int \frac{d^d k_1}{(2\pi)^d} \frac{d^d k_2}{(2\pi)^d} \frac{1}{((k_2)^2)^{a_1} ((k_1)^2)^{a_2} ((-p_2+k_1)^2)^{a_3}}$$

$$\frac{1}{((q_1+k_2)^2 - m_t^2)^{a_4} ((k_1+k_2)^2 - m_t^2)^{a_5} ((p_1-q_2+k_1+k_2)^2 - m_t^2)^{a_6}},$$ (B.1)

and there are 14 master integrals for this topology:

$$
\begin{aligned}
MI_{Sd1,1}(d) &= I(d,0,0,0,1,0,1), & MI_{Sd1,8}(d) &= I(d,0,2,0,1,1,1),\\
MI_{Sd1,2}(d) &= I(d,0,0,0,1,1,1), & MI_{Sd1,9}(d) &= I(d,1,0,1,1,1,1),\\
MI_{Sd1,3}(d) &= I(d,0,0,1,2,1,0), & MI_{Sd1,10}(d) &= I(d,1,1,1,1,1,1),\\
MI_{Sd1,4}(d) &= I(d,0,0,2,1,1,0), & MI_{Sd1,11}(d) &= I(d,1,2,0,1,0,1),\\
MI_{Sd1,5}(d) &= I(d,0,1,0,1,2,1), & MI_{Sd1,12}(d) &= I(d,2,0,1,0,1,1),\\
MI_{Sd1,6}(d) &= I(d,0,1,0,2,1,1), & MI_{Sd1,13}(d) &= I(d,2,0,1,1,1,1),\\
MI_{Sd1,7}(d) &= I(d,0,2,0,1,1,0), & MI_{Sd1,14}(d) &= I(d,2,1,0,1,0,1).
\end{aligned}
$$ (B.2)

The final formulas for the dimension shift of topology $Sd1$ are:

$$MI_{Sd1,1}(2+d) = (4\,m_t^4\,MI_{Sd1,1}(d))/d^2$$

$$MI_{Sd1,2}(2+d) = \left(m_t^2\,((2\,m_t^2-s-t)\,MI_{Sd1,1}(d) - (s+t)^2\,MI_{Sd1,2}(d))\right)/\left((-1+d)\,d\,(m_t^2-s-t)\right)$$

$$MI_{Sd1,3}(2+d) = -\Big((-3+d)\,m_t^2\,(16\,(15-16\,d+4\,d^2)\,m_t^4 - 4\,(80-87\,d+22\,d^2)\,m_t^2\,(s+t)$$

$$+ (88-98\,d+25\,d^2)\,(s+t)^2)\,MI_{Sd1,3}(d) + (-4+d)\,(s+t)\,(4\,(15-16\,d+4\,d^2)\,m_t^4$$

$$- 2\,(28-31\,d+8\,d^2)\,m_t^2\,(s+t) + (2-3\,d+d^2)\,(s+t)^2)\,MI_{Sd1,4}(d)\Big)/\Big(3\,(-3+d)$$

$$(64-168\,d+158\,d^2-63\,d^3+9\,d^4)\,(m_t^2-s-t)\Big)$$

B. Final formulas for the dimension shift of the topologies Sd1 and Sd2

$$MI_{Sd1,\,4}(2+d) = \Big(2\,m_t^2\,(8\,(-5+2\,d)\,m_t^4 - 8\,(-5+2\,d)\,m_t^2\,(s+t) + (-4+d)\,(s+t)^2)\,MI_{Sd1,\,3}(d)$$

$$- \Big[(-4+d)\,(s+t)\,((20-8\,d)\,m_t^4 + (-12+5\,d)\,m_t^2\,(s+t) + (-2+d)\,(s+t)^2)\,MI_{Sd1,\,4}(d)\Big]/(-3+d)\Big)$$

$$/\Big(3\,(-2+d)\,(-8+3\,d)\,(-4+3\,d)\,(m_t^2 - s - t)\Big)$$

$$MI_{Sd1,\,5}(2+d) = \Bigg(\Big[\Big(-64\,(-30+37\,d-15\,d^2+2\,d^3)\,m_t^{12} + 16\,(-340+416\,d-167\,d^2+22\,d^3)\,m_t^{10}\,(s+t)$$

$$- 4\,(-1440+1686\,d-649\,d^2+82\,d^3)\,m_t^8\,(s+t)^2 + 8\,(-240+250\,d-84\,d^2+9\,d^3)\,m_t^6\,(s+t)^3$$

$$+ (-464+644\,d-292\,d^2+43\,d^3)\,m_t^4\,(s+t)^4 - (-2+d)^2\,(-52+17\,d)\,m_t^2\,(s+t)^5$$

$$+ 2\,(-3+d)\,(-2+d)^2\,(s+t)^6\Big)\,MI_{Sd1,\,1}(d)\Big]/\Big((30-19\,d+3\,d^2)\,(s+t)^2\Big) + \Big[4\,m_t^2\,(m_t^2-s-t)$$

$$(32\,(15-11\,d+2\,d^2)\,m_t^{10} - 8\,(70-53\,d+10\,d^2)\,m_t^8\,(s+t) + 4\,(80-53\,d+9\,d^2)\,m_t^6\,(s+t)^2$$

$$- (80-30\,d+d^2)\,m_t^4\,(s+t)^3 - (56-46\,d+9\,d^2)\,m_t^2\,(s+t)^4 + 2\,(6-5\,d+d^2)\,(s+t)^5)\,MI_{Sd1,\,2}(d)\Big]$$

$$/\Big((-10+3\,d)\,(s+t)^2\Big) + \Big[8\,m_t^4\,(16\,(15-11\,d+2\,d^2)\,m_t^{10} - 4\,(140-101\,d+18\,d^2)\,m_t^8\,(s+t)$$

$$+ (320-214\,d+35\,d^2)\,m_t^6\,(s+t)^2 + 2\,(40-40\,d+9\,d^2)\,m_t^4\,(s+t)^3 - (68-56\,d+11\,d^2)\,m_t^2\,(s+t)^4$$

$$+ 2\,(6-5\,d+d^2)\,(s+t)^5)\,MI_{Sd1,\,5}(d)\Big]/(20-16\,d+3\,d^2) - \Big[4\,m_t^4\,(64\,(-105+107\,d-36\,d^2+4\,d^3)\,m_t^{12}$$

$$- 16\,(-900+915\,d-307\,d^2+34\,d^3)\,m_t^{10}\,(s+t) + 4\,(-2280+2266\,d-745\,d^2+81\,d^3)\,m_t^8\,(s+t)^2$$

$$- 2\,(-560+436\,d-110\,d^2+9\,d^3)\,m_t^6\,(s+t)^3 - (-704+808\,d-290\,d^2+33\,d^3)\,m_t^4\,(s+t)^4$$

$$+ (-320+344\,d-118\,d^2+13\,d^3)\,m_t^2\,(s+t)^5 - 2\,(-24+26\,d-9\,d^2+d^3)\,(s+t)^6)\,MI_{Sd1,\,6}(d)\Big]$$

$$/\Big((-60+68\,d-25\,d^2+3\,d^3)\,(s+t)\Big) + \Big[4\,m_t^4\,(32\,(35-24\,d+4\,d^2)\,m_t^{10}$$

$$- 8\,(265-181\,d+30\,d^2)\,m_t^8\,(s+t) + 6\,(148-96\,d+15\,d^2)\,m_t^6\,(s+t)^2$$

$$+ (176-158\,d+33\,d^2)\,m_t^4\,(s+t)^3 - (80-66\,d+13\,d^2)\,m_t^2\,(s+t)^4 + 2\,(6-5\,d+d^2)\,(s+t)^5)\,MI_{Sd1,\,7}(d)\Big]$$

$$/\Big((6-5\,d+d^2)\,(s+t)\Big) + \Big[4\,(-4+d)\,m_t^4\,(32\,(15-11\,d+2\,d^2)\,m_t^{10} - 64\,(15-11\,d+2\,d^2)\,m_t^8\,(s+t)$$

$$+ 20\,(24-17\,d+3\,d^2)\,m_t^6\,(s+t)^2 + (40-46\,d+11\,d^2)\,m_t^4\,(s+t)^3 - (56-46\,d+9\,d^2)\,m_t^2\,(s+t)^4$$

$$+ 2\,(6-5\,d+d^2)\,(s+t)^5)\,MI_{Sd1,\,8}(d)\Big]/(-60+68\,d-25\,d^2+3\,d^3)\Bigg)$$

$$/(24\,(-1+d)\,(-8+3\,d)\,m_t^6\,(-m_t^2+s+t)^2)$$

$$MI_{Sd1,\,6}(2+d) = \Bigg(\Big[(-2+d)\,(64\,(-30+37\,d-15\,d^2+2\,d^3)\,m_t^{12} - 16\,(-160+194\,d-77\,d^2+10\,d^3)\,m_t^{10}\,(s+t)$$

$$- 4\,(-816+900\,d-332\,d^2+41\,d^3)\,m_t^8\,(s+t)^2 + 4\,(-1068+1192\,d-441\,d^2+54\,d^3)\,m_t^6\,(s+t)^3$$

$$- 8\,(-40+46\,d-17\,d^2+2\,d^3)\,m_t^4\,(s+t)^4 - (-2+d)^2\,(-20+7\,d)\,m_t^2\,(s+t)^5$$

$$+ (-3+d)\,(-2+d)^2\,(s+t)^6)\,MI_{Sd1,\,1}(d)\Big]/\Big((30-19\,d+3\,d^2)\,(m_t^2 s - t)\Big)$$

$$- \Big[4\,(-2+d)\,m_t^2\,(32\,(15-11\,d+2\,d^2)\,m_t^{10} + 8\,(20-13\,d+2\,d^2)\,m_t^8\,(s+t)$$

$$- 2\,(248-182\,d+33\,d^2)\,m_t^6\,(s+t)^2 + (28-24\,d+5\,d^2)\,m_t^4\,(s+t)^3 + (16-14\,d+3\,d^2)\,m_t^2\,(s+t)^4$$

$$- (6-5\,d+d^2)\,(s+t)^5)\,MI_{Sd1,\,2}(d)\Big]/(-10+3\,d) - \Big[8\,m_t^4\,(s+t)^2\,(m_t^2+s+t)\,(16\,(15-11\,d+2\,d^2)\,m_t^8$$

$$- 4\,(110-79\,d+14\,d^2)\,m_t^6\,(s+t) + (172-116\,d+19\,d^2)\,m_t^4\,(s+t)^2 + (28-24\,d+5\,d^2)\,m_t^2\,(s+t)^3$$

$$- (6-5\,d+d^2)\,(s+t)^4)\,MI_{Sd1,\,5}(d)\Big]/\Big((-10+3\,d)\,(m_t^2-s-t)\Big)$$

$$+ \left[4\,m_t^4\,(s+t)\,(64\,(-105+107\,d-36\,d^2+4\,d^3)\,m_t^{12} - 16\,(-270+273\,d-91\,d^2+10\,d^3)\,m_t^{10}\,(s+t) \right.$$

$$-4\,(-1356+1376\,d-461\,d^2+51\,d^3)\,m_t^8\,(s+t)^2 + 4\,(-956+946\,d-308\,d^2+33\,d^3)\,m_t^6\,(s+t)^3$$

$$-2\,(-328+302\,d-91\,d^2+9\,d^3)\,m_t^4\,(s+t)^4 - (-112+124\,d-44\,d^2+5\,d^3)\,m_t^2\,(s+t)^5$$

$$\left. +(-24+26\,d-9\,d^2+d^3)\,(s+t)^6)\,MI_{Sd1,\,6}(d) \right] / \left((30-19\,d+3\,d^2)\,(m_t^2-s-t) \right)$$

$$- \left[4\,m_t^4\,(s+t)\,(32\,(35-24\,d+4\,d^2)\,m_t^{10} - 8\,(90-61\,d+10\,d^2)\,m_t^8\,(s+t) \right.$$

$$-2\,(292-204\,d+35\,d^2)\,m_t^6\,(s+t)^2 + 2\,(82-55\,d+9\,d^2)\,m_t^4\,(s+t)^3 + (28-24\,d+5\,d^2)\,m_t^2\,(s+t)^4$$

$$\left. - (6-5\,d+d^2)\,(s+t)^5)\,MI_{Sd1,\,7}(d) \right] / \left((-3+d)\,(m_t^2-s-t) \right) - \left[4\,(-4+d)\,m_t^4\,(s+t)^2\,(32\,(15 \right.$$

$$-11\,d+2\,d^2)\,m_t^{10} - 16\,(15-11\,d+2\,d^2)\,m_t^8\,(s+t) - 4\,(114-83\,d+15\,d^2)\,m_t^6\,(s+t)^2$$

$$+2\,(104-74\,d+13\,d^2)\,m_t^4\,(s+t)^3 + (16-14\,d+3\,d^2)\,m_t^2\,(s+t)^4$$

$$\left. - (6-5\,d+d^2)\,(s+t)^5)\,MI_{Sd1,\,8}(d) \right] / \left((30-19\,d+3\,d^2)\,(m_t^2-s-t) \right) \Bigg)$$

$$/ \left(12\,(-2+d)^2\,(-8+3\,d)\,m_t^6\,(s+t)^3 \right)$$

$$MI_{Sd1,\,7}(2+d) = (-16\,(-7+2\,d)\,(-5+2\,d)\,m_t^4\,MI_{Sd1,\,7}(d)) / (3\,(192-376\,d+266\,d^2-81\,d^3+9\,d^4))$$

$$MI_{Sd1,\,8}(2+d) = \left(-\left(\left[(-2+d)\,(32\,(-30+37\,d-15\,d^2+2\,d^3)\,m_t^{12} - 8\,(-196+242\,d-98\,d^2 \right. \right. \right.$$

$$+13\,d^3)\,m_t^{10}\,(s+t) - 4\,(-312+324\,d-112\,d^2+13\,d^3)\,m_t^8\,(s+t)^2 + 4\,(-528+590\,d-219\,d^2$$

$$+27\,d^3)\,m_t^6\,(s+t)^3 - (-280+340\,d-134\,d^2+17\,d^3)\,m_t^4\,(s+t)^4 + 2\,(-2+d)^2\,(-7+2\,d)\,m_t^2\,(s+t)^5$$

$$\left. \left. - (-3+d)\,(-2+d)^2\,(s+t)^6)\,MI_{Sd1,\,1}(d) \right] / ((30-19\,d+3\,d^2)\,(m_t^2-s-t)) \right)$$

$$+ \left[(4\,(-2+d)\,m_t^2\,(16\,(15-11\,d+2\,d^2)\,m_t^{10} - 4\,(-2-2\,d+d^2)\,m_t^8\,(s+t) \right.$$

$$-4\,(68-50\,d+9\,d^2)\,m_t^6\,(s+t)^2 + (80-66\,d+13\,d^2)\,m_t^4\,(s+t)^3$$

$$+2\,(-2+d)\,m_t^2\,(s+t)^4 + (6-5\,d+d^2)\,(s+t)^5)\,MI_{Sd1,\,2}(d) \right] / (-10+3\,d) + \left[8\,m_t^4\,(s+t)^2\,(m_t^2+s+t) \right.$$

$$(8\,(15-11\,d+2\,d^2)\,m_t^8 - 2\,(128-94\,d+17\,d^2)\,m_t^6\,(s+t) + 4\,(38-28\,d+5\,d^2)\,m_t^4\,(s+t)^2$$

$$\left. -2\,(8-6\,d+d^2)\,m_t^2\,(s+t)^3 + (6-5\,d+d^2)\,(s+t)^4)\,MI_{Sd1,\,5}(d) \right] / ((-10+3\,d)\,(m_t^2-s-t))$$

$$- \left[4\,m_t^4\,(s+t)\,(32\,(-105+107\,d-36\,d^2+4\,d^3)\,m_t^{12} - 16\,(-198+207\,d-71\,d^2+8\,d^3)\,m_t^{10}\,(s+t) \right.$$

$$-8\,(-258+254\,d-83\,d^2+9\,d^3)\,m_t^8\,(s+t)^2 + 2\,(-1208+1228\,d-410\,d^2+45\,d^3)\,m_t^6\,(s+t)^3$$

$$-(-640+632\,d-202\,d^2+21\,d^3)\,m_t^4\,(s+t)^4 + 2\,(-4+d)^2\,(-2+d)\,m_t^2\,(s+t)^5$$

$$\left. - (-24+26\,d-9\,d^2+d^3)\,(s+t)^6)\,MI_{Sd1,\,6}(d) \right] / ((30-19\,d+3\,d^2)\,(m_t^2-s-t)) + \left[4\,m_t^4\,(s+t) \right.$$

$$(16\,(35-24\,d+4\,d^2)\,m_t^{10} - 8\,(66-47\,d+8\,d^2)\,m_t^8\,(s+t) - 4\,(46-30\,d+5\,d^2)\,m_t^6\,(s+t)^2$$

$$\left. +(160-118\,d+21\,d^2)\,m_t^4\,(s+t)^3 - 2\,(8-6\,d+d^2)\,m_t^2\,(s+t)^4 + (6-5\,d+d^2)\,(s+t)^5)\,MI_{Sd1,\,7}(d) \right]$$

$$/((-3+d)\,(m_t^2-s-t)) + \left[4\,(-4+d)\,m_t^4\,(s+t)^2\,(16\,(15-11\,d+2\,d^2)\,m_t^{10} \right.$$

$$-4\,(48-37\,d+7\,d^2)\,m_t^8\,(s+t) - 8\,(24-17\,d+3\,d^2)\,m_t^6\,(s+t)^2 + (140-104\,d+19\,d^2)\,m_t^4\,(s+t)^3$$

$$\left. +2\,(-2+d)\,m_t^2\,(s+t)^4 + (6-5\,d+d^2)\,(s+t)^5)\,MI_{Sd1,\,8}(d) \right] / ((30-19\,d+3\,d^2)\,(m_t^2-s-t)) \Bigg)$$

$$/(12\,(-2+d)^2\,(-8+3\,d)\,m_t^6\,(s+t)^3)$$

B. Final formulas for the dimension shift of the topologies Sd1 and Sd2

$$
\begin{aligned}
MI_{Sd1,9}(2+d) = \Bigg(&\Big[3\,(-2+d)^2\,(2\,m_t^2 - s - t)\,(12\,(77 - 54\,d + 9\,d^2)\,m_t^4 - 12\,(77 - 54\,d + 9\,d^2)\,m_t^2\,(s+t) \\
&+ (-3+d)^2\,(-2+d)\,(s+t)^2)\,MI_{Sd1,1}(d) \Big] / ((20 - 9\,d + d^2)\,m_t^2\,(m_t^2 - s - t)\,(s+t)^2) \\
&- (8\,(-3+d)\,(-2+d)\,(36\,(56 - 45\,d + 9\,d^2)\,m_t^8 - 72\,(56 - 45\,d + 9\,d^2)\,m_t^6\,(s+t) \\
&+ 3\,(576 - 478\,d + 99\,d^2)\,m_t^4\,(s+t)^2 + 3\,(96 - 62\,d + 9\,d^2)\,m_t^2\,(s+t)^3 + 4\,(12 - 7\,d + d^2) \\
&\times\ (s+t)^4)\,MI_{Sd1,12}(d)) / ((80 - 54\,d + 9\,d^2)\,(-m_t^2 + s + t)^2) - \Big[48\,(-3+d)\,(-9+2\,d) \\
&\times\ m_t^4\,(12\,(-7+3\,d)\,m_t^4 - 12\,(-7+3\,d)\,m_t^2\,(s+t) + (-2+d)\,(s+t)^2)\,MI_{Sd1,13}(d)) \Big] \\
&/((-4+d)\,(m_t^2 - s - t)) - \Big[36\,(-3+d)\,(-2+d)\,(4\,(77 - 54\,d + 9\,d^2)\,m_t^4 \\
&- 4\,(77 - 54\,d + 9\,d^2)\,m_t^2\,(s+t) + (19 - 13\,d + 2\,d^2)\,(s+t)^2)\,MI_{Sd1,2}(d) \Big] / ((-5+d)\,(s+t)^2) \\
&+ \Big[8\,(-3+d)\,m_t^2\,(24\,(-63840 + 106852\,d - 70860\,d^2 + 23265\,d^3 - 3780\,d^4 + 243\,d^5)\,m_t^8 \\
&- 36\,(-27104 + 50228\,d - 36680\,d^2 + 13185\,d^3 - 2331\,d^4 + 162\,d^5)\,m_t^6\,(s+t) - 4\,(-599040 \\
&+ 926720\,d - 562288\,d^2 + 167040\,d^3 - 24264\,d^4 + 1377\,d^5)\,m_t^4\,(s+t)^2 + 2\,(-985056 + 1579084\,d \\
&- 999448\,d^2 + 312217\,d^3 - 48150\,d^4 + 2934\,d^5)\,m_t^2\,(s+t)^3 - (-135168 + 216208\,d - 136116\,d^2 \\
&+ 42176\,d^3 - 6435\,d^4 + 387\,d^5)\,(s+t)^4)\,MI_{Sd1,3}(d) \Big] / ((4480 - 5104\,d + 2148\,d^2 - 396\,d^3 \\
&+ 27\,d^4)\,(m_t^2 - s - t)\,(s+t)^3) + \Big[4\,(48\,(446880 - 875644\,d + 709724\,d^2 - 304575\,d^3 + 72990\,d^4 \\
&- 9261\,d^5 + 486\,d^6)\,m_t^{10} - 24\,(2181984 - 4293092\,d + 3494100\,d^2 - 1505685\,d^3 + 362286\,d^4 \\
&- 46143\,d^5 + 2430\,d^6)\,m_t^8\,(s+t) + 8\,(5312160 - 10477372\,d + 8548192\,d^2 - 3692341\,d^3 \\
&+ 890406\,d^4 - 113634\,d^5 + 5994\,d^6)\,m_t^6\,(s+t)^2 - 4\,(3068928 - 6037596\,d + 4912556\,d^2 \\
&- 2115913\,d^3 + 508731\,d^4 - 64719\,d^5 + 3402\,d^6)\,m_t^4\,(s+t)^3 + 4\,(165792 - 317588\,d + 251012\,d^2 \\
&- 104827\,d^3 + 24409\,d^4 - 3006\,d^5 + 153\,d^6)\,m_t^2\,(s+t)^4 + (-4+d)^2\,(840 - 1132\,d + 554\,d^2 - 117\,d^3 \\
&+ 9\,d^4)\,(s+t)^5)\,MI_{Sd1,4}(d) \Big] / ((4480 - 5104\,d + 2148\,d^2 - 396\,d^3 + 27\,d^4)\,(m_t^2 - s - t)\,(s+t)^3) \\
&- \Big[2\,(2\,m_t^2 - s - t)\,(36\,(6160 - 8478\,d + 4329\,d^2 - 972\,d^3 + 81\,d^4)\,m_t^8 - 72\,(6160 - 8478\,d \\
&+ 4329\,d^2 - 972\,d^3 + 81\,d^4)\,m_t^6\,(s+t) + (227712 - 313948\,d + 160568\,d^2 - 36107\,d^3 \\
&+ 3013\,d^4)\,m_t^4\,(s+t)^2 - (5952 - 8740\,d + 4724\,d^2 - 1115\,d^3 + 97\,d^4)\,m_t^2\,(s+t)^3 \\
&- 4\,(-3+d)^2\,(8 - 6\,d + d^2)\,(s+t)^4)\,MI_{Sd1,7}(d) \Big] / ((32 - 20\,d + 3\,d^2)\,(s+t)^2\,(-m_t^2 + s + t)^2) \\
&+ \Big[12\,(-3+d)\,m_t^2\,(48\,(77 - 54\,d + 9\,d^2)\,m_t^8 - 96\,(77 - 54\,d + 9\,d^2)\,m_t^6\,(s+t) + 24\,(160 - 113\,d \\
&+ 19\,d^2)\,m_t^4\,(s+t)^2 - 24\,(6 - 5\,d + d^2)\,m_t^2\,(s+t)^3 - (8 - 6\,d + d^2)\,(s+t)^4)\,MI_{Sd1,9}(d) \Big] \\
&/((m_t^2 - s - t)\,(s+t)^2) \Bigg) / (12\,(-3+d)^2\,(-2+d)^3)
\end{aligned}
$$

$$
\begin{aligned}
MI_{Sd1,10}(2+d) = \Bigg(&\Big[3\,(-2+d)\,(256\,(-3+d)^2\,(945 - 844\,d + 276\,d^2 - 39\,d^3 + 2\,d^4)\,m_t^{10} \\
&+ 16\,(239760 - 352912\,d + 205967\,d^2 - 59820\,d^3 + 8740\,d^4 - 540\,d^5 + 5\,d^6)\,m_t^8\,(s+t) \\
&- 8\,(480320 - 756382\,d + 484567\,d^2 - 160761\,d^3 + 28859\,d^4 - 2613\,d^5 + 90\,d^6)\,m_t^6\,(s+t)^2 \\
&+ 4\,(232960 - 377656\,d + 253972\,d^2 - 90567\,d^3 + 18019\,d^4 - 1889\,d^5 + 81\,d^6)\,m_t^4\,(s+t)^3
\end{aligned}
$$

$$- 2 \left(40320 - 56280\, d + 33060\, d^2 - 10670\, d^3 + 2043\, d^4 - 224\, d^5 + 11\, d^6\right) m_t^2 \left(s + t\right)^4$$

$$- \left(-5280 + 5276\, d - 1544\, d^2 - 65\, d^3 + 111\, d^4 - 19\, d^5 + d^6\right) \left(s + t\right)^5\right) MI_{Sd1,\,1}(d)\Big] /$$

$$\left((-3 + d)^2 \left(1600 - 1800\, d + 746\, d^2 - 135\, d^3 + 9\, d^4\right) m_t^6 \left(s + t\right)^3 \left(2\left(-5 + d\right) m_t^2 - \left(-4 + d\right)\left(s + t\right)\right)\right)$$

$$- \Big[24\left(-4 + d\right)\left(s + t\right)^2 MI_{Sd1,\,10}(d)\Big]/(-2 + d) + \Big[12\left(128\left(-3 + d\right)^2\left(14 - 11\, d + 2\, d^2\right) m_t^8 - 16\,(2520$$

$$- 3618\, d + 1913\, d^2 - 443\, d^3 + 38\, d^4\right) m_t^6 \left(s + t\right) + 4\left(2800 - 4610\, d + 2711\, d^2 - 684\, d^3$$

$$+ 63\, d^4\right) m_t^4 \left(s + t\right)^2 + 2\left(10 - 3\, d\right)^2\left(40 - 23\, d + 3\, d^2\right) m_t^2 \left(s + t\right)^3 + \left(480 - 724\, d + 404\, d^2 - 99\, d^3\right.$$

$$+ 9\, d^4\right)\left(s + t\right)^4\right) MI_{Sd1,\,11}(d)\Big]/\left((480 - 724\, d + 404\, d^2 - 99\, d^3 + 9\, d^4\right) m_t^2 \left(s + t\right)\left(2\left(-5 + d\right) m_t^2$$

$$- \left(-4 + d\right)\left(s + t\right)\right)) - \left(48\,(48\left(-112 + 130\, d - 49\, d^2 + 6\, d^3\right) m_t^6\right.$$

$$- 3\left(-2000 + 2358\, d - 899\, d^2 + 111\, d^3\right) m_t^4 \left(s + t\right) - 2\left(-944 + 818\, d - 223\, d^2 + 19\, d^3\right) m_t^2 \left(s + t\right)^2$$

$$+ \left(-272 + 374\, d - 151\, d^2 + 19\, d^3\right)\left(s + t\right)^3\right) MI_{Sd1,\,12}(d))/\left((480 - 724\, d + 404\, d^2 - 99\, d^3\right.$$

$$+ 9\, d^4\right)\left(m_t^2 - s - t\right)\left(-2\, m_t^2 + s + t\right)) - \Big[144\left(-9 + 2\, d\right) m_t^2 \left(32\left(-7 + 2\, d\right) m_t^4\right.$$

$$- 8\left(-7 + 2\, d\right) m_t^2 \left(s + t\right) - \left(-3 + d\right)\left(s + t\right)^2\right) MI_{Sd1,\,13}(d)\Big]/\left((-24 + 26\, d - 9\, d^2 + d^3\right)\left(-2\, m_t^2 + s + t\right))$$

$$- \Big[96\left(-5 + d\right)\left(-m_t^2 + s + t\right)^2 \left(64\left(7 - 2\, d\right)^2\left(-3 + d\right) m_t^6 - 8\left(-1400 + 1289\, d - 394\, d^2\right.$$

$$+ 40\, d^3\right) m_t^4 \left(s + t\right) + 4\left(-320 + 296\, d - 90\, d^2 + 9\, d^3\right) m_t^2 \left(s + t\right)^2 - \left(-240 + 242\, d - 81\, d^2\right.$$

$$+ 9\, d^3\right)\left(s + t\right)^3\right) MI_{Sd1,\,14}(d)\Big]/\left((640 - 912\, d + 476\, d^2 - 108\, d^3 + 9\, d^4\right)\left(s + t\right)^3\left(-2\left(-5 + d\right) m_t^2\right.$$

$$+ \left(-4 + d\right)\left(s + t\right)\right)) + \Big[24\left(m_t^2 - s - t\right)\left(64\left(-3 + d\right)^2\left(-140 + 103\, d - 25\, d^2 + 2\, d^3\right) m_t^8 - 4\,(-265200$$

$$+ 395098\, d - 234275\, d^2 + 69143\, d^3 - 10161\, d^4 + 595\, d^5\right) m_t^6 \left(s + t\right) + 2\left(-174480 + 257290\, d\right.$$

$$- 150773\, d^2 + 43919\, d^3 - 6363\, d^4 + 367\, d^5\right) m_t^4 \left(s + t\right)^2 + \left(-8160 + 10676\, d - 5682\, d^2 + 1553\, d^3\right.$$

$$- 220\, d^4 + 13\, d^5\right) m_t^2 \left(s + t\right)^3 + \left(-240 + 1118\, d - 1069\, d^2 + 421\, d^3 - 75\, d^4 + 5\, d^5\right)\left(s + t\right)^4\right) MI_{Sd1,\,2}(d)\Big]$$

$$/\left((-4800 + 7000\, d - 4038\, d^2 + 1151\, d^3 - 162\, d^4 + 9\, d^5\right) m_t^4 \left(s + t\right)^3\left(-2\, m_t^2 + s + t\right))$$

$$+ \Big[8\left(1536\left(7 - 2\, d\right)^2\left(-3 + d\right)^3\left(-14 + 3\, d\right) m_t^{12} + 192\left(190344 - 293854\, d + 181589\, d^2 - 56568\, d^3\right.$$

$$+ 9073\, d^4 - 656\, d^5 + 12\, d^6\right) m_t^{10} \left(s + t\right) + 96\left(530208 - 873794\, d + 599393\, d^2 - 219744\, d^3 + 45539\, d^4\right.$$

$$- 5069\, d^5 + 237\, d^6\right) m_t^8 \left(s + t\right)^2 - 8\left(30152544 - 50683944\, d + 35211358\, d^2 - 12946157\, d^3 + 2657607\, d^4\right.$$

$$- 288855\, d^5 + 12987\, d^6\right) m_t^6 \left(s + t\right)^3 + 4\left(38130624 - 65039256\, d + 45896644\, d^2 - 17158734\, d^3\right.$$

$$+ 3585585\, d^4 - 397188\, d^5 + 18225\, d^6\right) m_t^4 \left(s + t\right)^4 - 2\left(13811520 - 23913720\, d + 17145248\, d^2 - 6518100\, d^3\right.$$

$$+ 1386255\, d^4 - 156420\, d^5 + 7317\, d^6\right) m_t^2 \left(s + t\right)^5 + \left(-3 + d\right)^2\left(46400 - 53720\, d + 22980\, d^2 - 4302\, d^3\right.$$

$$+ 297\, d^4\right)\left(s + t\right)^6\right) MI_{Sd1,\,3}(d)\Big]/\left((26880 - 53024\, d + 42888\, d^2 - 18220\, d^3 + 4290\, d^4 - 531\, d^5 + 27\, d^6\right)$$

$$\times \left(s + t\right)^4\left(-2\, m_t^2 + s + t\right)\left(2\left(-5 + d\right) m_t^2 - \left(-4 + d\right)\left(s + t\right)\right)) + \Big[4\left(3072\left(-14 + 3\, d\right)\left(21 - 13\, d + 2\, d^2\right)^3 m_t^{14}\right.$$

$$- 384\left(7 - 2\, d\right)^2\left(7032 - 5362\, d - 233\, d^2 + 1166\, d^3 - 341\, d^4 + 30\, d^5\right) m_t^{12} \left(s + t\right) - 192\,(-5948544$$

$$+ 11608546\, d - 9573777\, d^2 + 4318454\, d^3 - 1148126\, d^4 + 179378\, d^5 - 15185\, d^6 + 534\, d^7\right) m_t^{10} \left(s + t\right)^2$$

$$+ 16\left(-46912992 + 93131544\, d - 78340886\, d^2 + 36172349\, d^3 - 9892691\, d^4 + 1600863\, d^5 - 141759\, d^6\right.$$

$$+ 5292\, d^7\right) m_t^8 \left(s + t\right)^3 - 24\left(-5594784 + 10038212\, d - 7464756\, d^2 + 2951965\, d^3 - 657962\, d^4 + 79466\, d^5\right.$$

$$- 4326\, d^6 + 45\, d^7\right) m_t^6 \left(s + t\right)^4 - 4\left(-3076800 + 11591208\, d - 14863700\, d^2 + 9498828\, d^3 - 3411546\, d^4\right.$$

$$+ 702837\, d^5 - 77778\, d^6 + 3591\, d^7\right) m_t^4 \left(s + t\right)^5 + 2\left(-4029120 + 10400616\, d - 11056404\, d^2 + 6329782\, d^3\right.$$

$$-\, 2120029\, d^4 + 417087\, d^5 - 44757\, d^6 + 2025\, d^7)\, m_t^2\, (s+t)^6 - 3\, (30 - 19\, d + 3\, d^2)^2\, (-224 + 244\, d$$

$$-\, 84\, d^2 + 9\, d^3)\, (s+t)^7)\, MI_{Sd1,\,4}(d)\Big]\big/((-3+d)^2\, (-8960 + 14688\, d - 9400\, d^2 + 2940\, d^3 - 450\, d^4$$

$$+\, 27\, d^5)\, m_t^2\, (s+t)^4\, (-2\, m_t^2 + s + t)\, (2\, (-5+d)\, m_t^2 - (-4+d)\, (s+t))) - \Big[48\, (m_t^2 + s + t)\, (16\, (-84$$

$$+\, 73\, d - 21\, d^2 + 2\, d^3)\, m_t^4 + (208 - 54\, d - 19\, d^2 + 5\, d^3)\, m_t^2\, (s+t) - (-16 + 66\, d - 35\, d^2 + 5\, d^3)$$

$$\times\, (s+t)^2)\, MI_{Sd1,\,5}(d)\Big]\big/((640 - 912\, d + 476\, d^2 - 108\, d^3 + 9\, d^4)\, m_t^2\, (s+t)) + \Big[24\, (64\, (7 - 2\, d)^2$$

$$\times\, (-3+d)\, m_t^8 + 8\, (-1008 + 967\, d - 306\, d^2 + 32\, d^3)\, m_t^6\, (s+t) - 4\, (-4+d)^2\, (-29 + 9\, d)\, m_t^4\, (s+t)^2$$

$$+\, (-688 + 538\, d - 143\, d^2 + 13\, d^3)\, m_t^2\, (s+t)^3 + (-16 + 66\, d - 35\, d^2 + 5\, d^3)\, (s+t)^4)\, MI_{Sd1,\,6}(d)\Big]$$

$$\big/((480 - 724\, d + 404\, d^2 - 99\, d^3 + 9\, d^4)\, m_t^2\, (s+t)^2) + \Big[12\, (16\, (16128 - 20078\, d + 9313\, d^2 - 1909\, d^3$$

$$+\, 146\, d^4)\, m_t^8 - 4\, (75376 - 93706\, d + 43447\, d^2 - 8909\, d^3 + 682\, d^4)\, m_t^6\, (s+t) + (49360 - 60374\, d$$

$$+\, 27677\, d^2 - 5632\, d^3 + 429\, d^4)\, m_t^4\, (s+t)^2 + (912 - 2158\, d + 1461\, d^2 - 392\, d^3 + 37\, d^4)\, m_t^2\, (s+t)^3$$

$$-\, 2\, (48 - 214\, d + 171\, d^2 - 50\, d^3 + 5\, d^4)\, (s+t)^4)\, MI_{Sd1,\,7}(d)\Big]\big/((-3+d)^2\, (-64 + 72\, d - 26\, d^2 + 3\, d^3)$$

$$\times\, m_t^2\, (m_t^2 - s - t)\, (s+t)^2) - \Big[24\, (32\, (-84 + 73\, d - 21\, d^2 + 2\, d^3)\, m_t^6 + 12\, (-224 + 204\, d - 61\, d^2 + 6\, d^3)$$

$$\times\, m_t^4\, (s+t) - (-1200 + 1154\, d - 375\, d^2 + 41\, d^3)\, m_t^2\, (s+t)^2 - (-16 + 66\, d - 35\, d^2 + 5\, d^3)\, (s+t)^3)$$

$$\times\, MI_{Sd1,\,8}(d)\Big]\big/((480 - 724\, d + 404\, d^2 - 99\, d^3 + 9\, d^4)\, m_t^2\, (s+t)) + \Big[48\, (96\, (77 - 43\, d + 6\, d^2)\, m_t^8$$

$$-\, 120\, (77 - 43\, d + 6\, d^2)\, m_t^6\, (s+t) + 4\, (457 - 260\, d + 37\, d^2)\, m_t^4\, (s+t)^2 + (148 - 81\, d + 11\, d^2)$$

$$\times\, m_t^2\, (s+t)^3 + (12 - 7\, d + d^2)\, (s+t)^4)\, MI_{Sd1,\,9}(d)\Big]\big/((6 - 5\, d + d^2)\, (s+t)^2\, (-2\, m_t^2 + s + t))\Bigg)$$

$$\big/(96\, (-7 + 2\, d)\, (-5 + 2\, d))$$

$$MI_{Sd1,\,11}(2+d) = \Bigg(\Big[-2\, (-2+d)\, (s+t)\, (-8\, (210 - 319\, d + 179\, d^2 - 44\, d^3 + 4\, d^4)\, m_t^6 + 4\, (1424 - 2106\, d$$

$$+\, 1147\, d^2 - 273\, d^3 + 24\, d^4)\, m_t^4\, (s+t) - 2\, (3868 - 5318\, d + 2714\, d^2 - 610\, d^3 + 51\, d^4)\, m_t^2\, (s$$

$$+\, t)^2 + (2328 - 3200\, d + 1636\, d^2 - 369\, d^3 + 31\, d^4)\, (s+t)^3)\, MI_{Sd1,\,1}(d)\Big]\big/(12 - 7\, d + d^2) - 4\, (-2$$

$$+\, d)\, m_t^4\, (s+t)^3\, (8\, (15 - 11\, d + 2\, d^2)\, m_t^4 - 2\, (142 - 99\, d + 17\, d^2)\, m_t^2\, (s+t) + (188 - 120\, d + 19\, d^2)$$

$$\times\, (s+t)^2)\, MI_{Sd1,\,11}(d) - \Big[32\, (-5+d)\, (-3+d)\, m_t^4\, (s+t)\, (-m_t^2 + s + t)^2\, (4\, (35 - 24\, d + 4\, d^2)\, m_t^4$$

$$-\, 2\, (74 - 49\, d + 8\, d^2)\, m_t^2\, (s+t) + (-4+d)^2\, (s+t)^2)\, MI_{Sd1,\,14}(d)\Big]\big/(-4+d) + \Big[2\, m_t^2\, (64\, (-3+d)^2\, (35$$

$$-\, 24\, d + 4\, d^2)\, m_t^{10} - 32\, (1506 - 2056\, d + 1049\, d^2 - 237\, d^3 + 20\, d^4)\, m_t^8\, (s+t) + 16\, (1970 - 2837\, d$$

$$+\, 1516\, d^2 - 356\, d^3 + 31\, d^4)\, m_t^6\, (s+t)^2 - 4\, (808 - 1660\, d + 1082\, d^2 - 287\, d^3 + 27\, d^4)\, m_t^4\, (s+t)^3$$

$$-\, 2\, (-4+d)^2\, (130 - 79\, d + 12\, d^2)\, m_t^2\, (s+t)^4 + (-4+d)^3\, (-10 + 3\, d)\, (s+t)^5)\, MI_{Sd1,\,3}(d)\Big]\big/(-4+d)$$

$$+\, \Big[(128\, (-5 + 2\, d)\, (21 - 13\, d + 2\, d^2)^2\, m_t^{12} - 128\, (-3+d)^2\, (-819 + 787\, d - 249\, d^2 + 26\, d^3)\, m_t^{10}\, (s+t)$$

$$+\, 32\, (-35970 + 57983\, d - 37285\, d^2 + 11956\, d^3 - 1912\, d^4 + 122\, d^5)\, m_t^8\, (s+t)^2 - 8\, (-75136$$

$$+\, 119120\, d - 75506\, d^2 + 23923\, d^3 - 3789\, d^4 + 240\, d^5)\, m_t^6\, (s+t)^3 + 4\, (-26960 + 41768\, d - 26014\, d^2$$

$$+\, 8143\, d^3 - 1281\, d^4 + 81\, d^5)\, m_t^4\, (s+t)^4 - 4\, (-4+d)^2\, (-70 + 71\, d - 25\, d^2 + 3\, d^3)\, m_t^2\, (s+t)^5$$

$$-\, (-4+d)^3\, (20 - 16\, d + 3\, d^2)\, (s+t)^6)\, MI_{Sd1,\,4}(d)\Big]\big/(12 - 7\, d + d^2)\Big)\big/(3\, (-2+d)^2\, (80 - 54\, d$$

$$+\, 9\, d^2)\, (s+t)^4\, (2\, (-5+d)\, m_t^2 - (-4+d)\, (s+t))$$

$$MI_{Sd1,\,12}(2+d) = -\Big(\big(4\,(-4+d)\,(s+t)^4\,MI_{Sd1,\,12}(d)\big)/(-10+3\,d) + \big[(2\,m_t^2 - s - t)\,(2\,(-7+2\,d)\,m_t^4$$
$$-\,2\,(-7+2\,d)\,m_t^2\,(s+t) - (-3+d)\,(s+t)^2)\,MI_{Sd1,\,7}(d)\big]/(-3+d)\Big)/(3\,(-2+d)\,(-8$$
$$+\,3\,d)\,(-m_t^2 + s + t)^2)$$

$$MI_{Sd1,\,13}(2+d) = \bigg(\Big(-3\,(-2+d)\,(-11+3\,d)\,(2\,m_t^2 - s - t)\,MI_{Sd1,\,1}(d)\Big)/((20 - 9\,d + d^2)\,m_t^2\,(s+t)^2)$$

$$+\,\big[8\,(-3+d)\,(3\,(-8+3\,d)\,m_t^4 - 3\,(-8+3\,d)\,m_t^2\,(s+t) - (-4+d)\,(s+t)^2)\,MI_{Sd1,\,12}(d)\big]/((80$$
$$-\,54\,d + 9\,d^2)\,(m_t^2 - s - t)) + \big[48\,(-3+d)\,(-9+2\,d)\,m_t^4\,MI_{Sd1,\,13}(d)\big]/(8 - 6\,d + d^2) + \big[(-3$$
$$+\,d)\,(12\,(-11+3\,d)\,m_t^4 - 12\,(-11+3\,d)\,m_t^2\,(s+t) + (-5+d)\,(s+t)^2)\,MI_{Sd1,\,2}(d)\big]/((-5+d)\,m_t^2$$
$$\times\,(s+t)^2) - \big[4\,(-3+d)\,(4\,(9120 - 11356\,d + 5256\,d^2 - 1071\,d^3 + 81\,d^4)\,m_t^6 - 2\,(-6624 + 6164\,d$$
$$-\,1884\,d^2 + 189\,d^3)\,m_t^4\,(s+t) - (44672 - 52624\,d + 22988\,d^2 - 4416\,d^3 + 315\,d^4)\,m_t^2\,(s+t)^2$$
$$+\,(1680 - 1844\,d + 752\,d^2 - 135\,d^3 + 9\,d^4)\,(s+t)^3)\,MI_{Sd1,\,3}(d)\big]/((-8960 + 14688\,d - 9400\,d^2$$
$$+\,2940\,d^3 - 450\,d^4 + 27\,d^5)\,(s+t)^3) + \big[2\,(-8\,(-63840 + 97732\,d - 59504\,d^2 + 18009\,d^3$$
$$-\,2709\,d^4 + 162\,d^5)\,m_t^6 + 4\,(-184032 + 284244\,d - 174560\,d^2 + 53265\,d^3 - 8073\,d^4 + 486\,d^5)$$
$$\times\,m_t^4\,(s+t) - 2\,(-131776 + 202792\,d - 124124\,d^2 + 37754\,d^3 - 5703\,d^4 + 342\,d^5)\,m_t^2\,(s+t)^2$$
$$+\,(-4+d)^2\,(-700 + 640\,d - 189\,d^2 + 18\,d^3)\,(s+t)^3)\,MI_{Sd1,\,4}(d)\big]/((-8960 + 14688\,d - 9400\,d^2$$
$$+\,2940\,d^3 - 450\,d^4 + 27\,d^5)\,(s+t)^3) + \big[2\,(2\,m_t^2 - s - t)\,(3\,(-880 + 834\,d - 261\,d^2 + 27\,d^3)\,m_t^4$$
$$-\,3\,(-880 + 834\,d - 261\,d^2 + 27\,d^3)\,m_t^2\,(s+t) + (-24 + 26\,d - 9\,d^2 + d^3)\,(s+t)^2)\,MI_{Sd1,\,7}(d)\big]$$
$$/((-64 + 72\,d - 26\,d^2 + 3\,d^3)\,(m_t^2 - s - t)\,(s+t)^2) - \big[4\,(-3+d)\,m_t^2\,(12\,(-11+3\,d)\,m_t^4$$
$$-\,12\,(-11+3\,d)\,m_t^2\,(s+t) + (-2+d)\,(s+t)^2)\,MI_{Sd1,\,9}(d)\big]/((-2+d)\,(s+t)^2)\bigg)/(2\,(-3+d)^2)$$

$$MI_{Sd1,\,14}(2+d) = \bigg(-\Big(\big[(16\,(210 - 319\,d + 179\,d^2 - 44\,d^3 + 4\,d^4)\,m_t^6 - 4\,(1840 - 2706\,d + 1463\,d^2$$
$$-\,345\,d^3 + 30\,d^4)\,m_t^4\,(s+t) + 2\,(920 - 1552\,d + 916\,d^2 - 230\,d^3 + 21\,d^4)\,m_t^2\,(s+t)^2 + (240$$
$$-\,236\,d + 88\,d^2 - 15\,d^3 + d^4)\,(s+t)^3)\,MI_{Sd1,\,1}(d)\big]/((12 - 7\,d + d^2)\,m_t^2\,(s+t)^2\,(2\,(-5+d)\,m_t^2$$
$$-\,(-4+d)\,(s+t)))\Big) + \big[4\,m_t^2\,(8\,(15 - 11\,d + 2\,d^2)\,m_t^4 - 4\,(35 - 24\,d + 4\,d^2)\,m_t^2\,(s+t)$$
$$+\,(20 - 9\,d + d^2)\,(s+t)^2)\,MI_{Sd1,\,11}(d)\big]/(2\,(-5+d)\,m_t^2 - (-4+d)\,(s+t))$$
$$-\,\big[4\,(-5+d)\,m_t^2\,(m_t^2 - s - t)\,(32\,(-105 + 107\,d - 36\,d^2 + 4\,d^3)\,m_t^6$$
$$-\,8\,(-360 + 369\,d - 125\,d^2 + 14\,d^3)\,m_t^4\,(s+t) + 10\,(-4+d)^2\,(-3+d)\,m_t^2\,(s+t)^2$$
$$+\,(-4+d)^2\,d\,(s+t)^3)\,MI_{Sd1,\,14}(d)\big]/((8 - 6\,d + d^2)\,(s+t)^2\,(-2\,(-5+d)\,m_t^2 + (-4+d)\,(s+t)))$$
$$+\,\big[2\,m_t^2\,(64\,(-3+d)^2\,(35 - 24\,d + 4\,d^2)\,m_t^8 - 16\,(1500 - 2105\,d + 1102\,d^2 - 255\,d^3$$
$$+\,22\,d^4)\,m_t^6\,(s+t) + 4\,(680 - 1424\,d + 955\,d^2 - 260\,d^3 + 25\,d^4)\,m_t^4\,(s+t)^2$$
$$+\,12\,(-4+d)^2\,(15 - 11\,d + 2\,d^2)\,m_t^2\,(s+t)^3 - (-4+d)^2\,(20 - 16\,d$$
$$+\,3\,d^2)\,(s+t)^4)\,MI_{Sd1,\,3}(d)\big]/\Big((8 - 6\,d + d^2)\,(s+t)^3\,(-2\,(-5+d)\,m_t^2 + (-4+d)\,(s+t))\Big)$$

$$+ \Big[(128\,(-5+2\,d)\,(21 - 13\,d + 2\,d^2)^2\,m_t^{10} - 32\,(-3+d)^2\,(-2100 + 2035\,d - 648\,d^2$$
$$+ 68\,d^3)\,m_t^8\,(s+t) + 8\,(-53160 + 85808\,d - 55261\,d^2 + 17752\,d^3 - 2845\,d^4 + 182\,d^5)\,m_t^6\,(s+t)^2$$
$$- 4\,(-26720 + 40708\,d - 24952\,d^2 + 7700\,d^3 - 1197\,d^4 + 75\,d^5)\,m_t^4\,(s+t)^3 + 4\,(-4+d)^2\,(-130$$
$$+ 109\,d - 31\,d^2 + 3\,d^3)\,m_t^2\,(s+t)^4 + (-10 + 3\,d)\,(8 - 6\,d + d^2)^2\,(s+t)^5)\,MI_{Sd1,\,4}(d) \Big]$$
$$\Big/ \Big((-24 + 26\,d - 9\,d^2 + d^3)\,(s+t)^3\,(-2\,(-5+d)\,m_t^2 + (-4+d)\,(s+t)) \Big) \Big)$$
$$\Big/ \Big(3\,(-1+d)\,(-10 + 3\,d)\,(-8 + 3\,d) \Big),$$

where t and s are the Mandelstam variables

$$p_1 \cdot p_2 = \frac{s}{2}, \qquad p_2 \cdot q_2 = -\frac{t}{2}, \tag{B.3}$$

m_t the top mass and d the space-time dimension.

The scalar integral belonging to topology $Sd2$ is:

$$I_{Sd2}(d, a_1, a_2, a_3, a_4, a_5, a_6) = \int \frac{d^d k_1}{(2\pi)^d}\frac{d^d k_2}{(2\pi)^d} \frac{1}{((k_2)^2)^{a_1}\,((k_1)^2)^{a_2}\,((-p_1+k_1)^2)^{a_3}}$$
$$\frac{1}{((q_2+k_2)^2)^{a_4}\,((k_1+k_2)^2)^{a_5}\,((p_2 - q1 + k_1 + k_2)^2)^{a_6}},$$

and there are 5 master integrals for this topology:

$$MI_{Sd2,\,1}(d) = I(d, 0, 0, 2, 1, 1, 0), \qquad MI_{Sd2,\,4}(d) = I(d, 2, 0, 1, 0, 1, 1),$$
$$MI_{Sd2,\,2}(d) = I(d, 0, 2, 0, 1, 1, 1), \qquad MI_{Sd2,\,5}(d) = I(d, 2, 1, 0, 0, 0, 1),$$
$$MI_{Sd2,\,3}(d) = I(d, 1, 1, 1, 1, 1, 1). \tag{B.4}$$

The final formulas for the dimension shift of the topology $Sd1$ are:

$$MI_{Sd2,\,1}(2+d) = \Big((-4+d)\,(mt^2 - s - t)^2\,MI_{Sd2,\,1}(d) \Big) \Big/ \Big(3\,(-3+d)\,(-8+3\,d)\,(-4+3\,d) \Big)$$
$$MI_{Sd2,\,2}(2+d) = \Big(-4\,(-4+d)\,(mt^2 - s - t)^2\,MI_{Sd2,\,2}(d) \Big) \Big/ \Big(3\,(-2+d)\,(-10+3\,d)\,(-8+3\,d) \Big)$$
$$MI_{Sd2,\,3}(2+d) = -\Big((2352 - 3530\,d + 1963\,d^2 - 478\,d^3 + 43\,d^4)\,MI_{Sd2,\,1}(d) \Big)$$
$$\Big/ \Big(4\,(-4+d)\,(-3+d)^2\,(-2+d)\,(-7+2\,d)\,(-5+2\,d)\,(-8+3\,d) \Big)$$
$$- \Big((-752 + 858\,d - 313\,d^2 + 37\,d^3)\,(mt^2 - s - t)\,MI_{Sd2,\,2}(d) \Big)$$
$$\Big/ \Big(2\,(-3+d)\,(-2+d)\,(-7+2\,d)\,(-5+2\,d)\,(-10+3\,d)\,(-8+3\,d) \Big)$$
$$- \Big((-4+d)\,(mt^2 - s - t)^2\,MI_{Sd2,\,3}(d) \Big) \Big/ \Big(4\,(-2+d)\,(-7+2\,d)\,(-5+2\,d) \Big)$$
$$- \Big((-752 + 858\,d - 313\,d^2 + 37\,d^3)\,(mt^2 - s - t)\,MI_{Sd2,\,4}(d) \Big)$$
$$\Big/ \Big(2\,(-3+d)\,(-2+d)\,(-7+2\,d)\,(-5+2\,d)\,(-10+3\,d)\,(-8+3\,d) \Big)$$

$$- \Big((2352 - 3530\,d + 1963\,d^2 - 478\,d^3 + 43\,d^4)\, MI_{Sd2,\,5}(d) \Big)$$

$$/ \Big(4\,(-4+d)\,(-3+d)^2\,(-2+d)\,(-7+2\,d)\,(-5+2\,d)\,(-8+3\,d) \Big)$$

$$MI_{Sd2,\,4}(2+d) = \Big(-4\,(-4+d)\,(mt^2 - s - t)^2\, MI_{Sd2,\,4}(d) \Big) / \Big(3\,(-2+d)\,(-10+3\,d)\,(-8+3\,d) \Big)$$

$$MI_{Sd2,\,5}(2+d) = \Big((-4+d)\,(mt^2 - s - t)^2\, MI_{Sd2,\,5}(d) \Big) / \Big(3\,(-3+d)\,(-8+3\,d)\,(-4+3\,d) \Big). \tag{B.5}$$

Single top quark production final result: Vertex contributions

On the next pages, we present the final result of the vertex contributions to single top quark productions, while m_t is set equal to 1.

t and s are the Mandelstam variables

$$p_1 \cdot p_2 = \frac{s}{2}, \qquad p_2 \cdot q_2 = -\frac{t}{2}, \tag{C.1}$$

and d the space-time dimension. C_F and C_A are the Casimir-Operators and N denotes the dimension of the fundamental representation of $SU(N)$. $I_2(R)$ sets the normalization of the representation R. n_f is the number of massles quark flavors up to the *bottom* quark. The occuring master integrals are:

$$
\begin{aligned}
master[MI101MI101] &= I_{Sd1}(d,0,0,0,1,0,1,0,0,0) \\
master[MI101MI202] &= I_{Sd1}(d,0,0,0,1,1,1,0,0,0) \\
master[MI201MI202] &= {}^{1}I_{dia65}(d,0,0,1,1,1,1,0,0,0) \\
master[MI301] &= I_{Sd2}(d,0,0,1,1,1,0,0,0,0) \\
master[MI302p] &= I_{Sd1}(d,0,0,2,1,1,0,0,0,0) \\
master[MI401] &= I_{Sd2}(d,1,0,1,0,1,1,0,0,0) \\
master[MI403p] &= I_{Sd1}(d,1,1,0,1,0,1,0,0,0) \\
master[MI406t] &= I_{v5}(d,1,1,0,1,1,0,0,0,0) \\
master[MI407t] &= I_{v5}(d,2,1,0,1,1,0,0,0,0) \\
master[MI409p] &= I_{Sd1}(d,0,1,0,1,1,1,0,0,0) \\
master[MI411p] &= I_{Sd1}(d,0,2,0,1,1,1,0,0,0) \\
master[MI502p] &= I_{Sd1}(d,2,0,1,1,1,1,0,0,0) \\
master[MI601p] &= I_{Sd1}(d,1,1,1,1,1,1,0,0,0) \\
master[MI301pu1] &= I_{Sd1}(d,0,1,0,1,1,0,0,0,0) \\[4pt]
master[MI101MI201] &= I_{v5}(d,1,0,0,1,1,0,0,0,0) \\
master[MI201MI201] &= I_{Sd8}(d,0,1,1,1,0,1,0,0,0) \\
master[MI202MI202] &= I_{Sd7}(d,0,1,1,1,1,0,0,0,0) \\
master[MI301p] &= I_{v2}(d,0,1,1,0,0,1,0,0,0) \\
master[MI303p] &= I_{v2}(d,1,0,0,0,1,1,0,0,0) \\
master[MI402p] &= I_{Sd1}(d,1,0,1,0,1,1,0,0,0) \\
master[MI404p] &= I_{Sd1}(d,2,1,0,1,0,1,0,0,0) \\
master[MI407p] &= I_{v2}(d,1,1,0,0,1,1,0,0,0) \\
master[MI408p] &= I_{v2}(d,1,2,0,0,1,1,0,0,0) \\
master[MI410p] &= I_{Sd1}(d,0,1,0,1,2,1,0,0,0) \\
master[MI501p] &= I_{Sd1}(d,1,0,1,1,1,1,0,0,0) \\
master[MI601] &= I_{Sd2}(d,1,1,1,1,1,1,0,0,0) \\
master[MI302pu1] &= I_{Sd7}(d,2,0,0,0,1,1,0,0,0)
\end{aligned}
\tag{C.2}
$$

---

[1] See eq.(5.36).

C. *Single top quark production final result: Vertex contributions*

$$\sum_{\text{vertex diagrams}} \mathcal{A}_i =$$

$S1\,((C_F\,(-2+d)\,N^2\,((4\,(-C_A+2\,C_F)\,(-231744+213488\,d+49536\,d^2-144902\,d^3+81307\,d^4-22784\,d^5+3541\,d^6-292\,d^7+10$

$\times\,d^8-4\,(-316736+694414\,d-627891\,d^2+302629\,d^3-82950\,d^4+12443\,d^5-799\,d^6-13\,d^7+3\,d^8)\,t-2\,(631872-1924268$

$\times\,d+2329726\,d^2-1502104\,d^3+569691\,d^4-130247\,d^5+17391\,d^6-1213\,d^7+32\,d^8)\,t^2-2\,(267648-745868\,d+840998$

$\times\,d^2-514630\,d^3+189876\,d^4-43753\,d^5+6214\,d^6-503\,d^7+18\,d^8)\,t^3+(-608320+1266968\,d-1102980\,d^2+516710\,d^3-138421$

$\times\,d^4+20242\,d^5-1207\,d^6-38\,d^7+6\,d^8)\,t^4-2\,(-4+d)^2\,(-2400+4100\,d-2744\,d^2+899\,d^3-144\,d^4+9\,d^5)\,t^5))/((12-7$

$\times\,d+d^2)^2\,(40-23\,d+3\,d^2)\,(1-t)^3\,(1+t)\,(-6+d+10\,t-3\,d\,t))-(8\,C_F\,(-2+d)\,(-1+d)\,(170-12\,d^2\,(-3+t)+d^3\,(-3+t)-62$

$\times\,t+d\,(-139+47\,t)))/((-4+d)\,(-3+d)^2\,(-1+t)^2)-(8\,C_A\,(2\,d^4\,(-1+t)^2-9\,d^3\,(3-12\,t+t^2)-3\,d^2\,(-59+232\,t+11\,t^2)-12$

$\times\,(-56+101\,t+21\,t^2)+4\,d\,(-142+407\,t+53\,t^2)))/((-3+d)^2\,(20-9\,d+d^2)\,(-1+t)^3)-(16\,I_2(R)\,(d^2\,(161352+471252$

$\times\,t-127572\,t^2)+d^4\,(24884+64624\,t-18932\,t^2)+48\,(1699+5017\,t-1226\,t^2)+d^6\,(424+932\,t-292\,t^2)+d^7\,(-17-34\,t+11$

$\times\,t^2)+2\,d^5\,(-2207-5294\,t+1605\,t^2)+12\,d\,(-14567-43538\,t+11233\,t^2)+d^3\,(-82377-228946\,t+64627\,t^2)))/((-5+d)^2$

$\times\,(-576+1224\,d-914\,d^2+313\,d^3-50\,d^4+3\,d^5)\,(-1+t)^3)-(4\,C_F\,(756+d^5\,(-1+t)^3-6468\,t+10672\,t^2+56\,t^3-2\,d^4\,(-10+7$

$\times\,t-76\,t^2+7\,t^3)+12\,d\,(-87+461\,t-1178\,t^2+8\,t^3)+d^3\,(-153+135\,t-1591\,t^2+73\,t^3)-2\,d^2\,(-285+740\,t-3499\,t^2+80$

$\times\,t^3)))/((-3+d)^2\,(20-9\,d+d^2)\,(-1+t)^3\,t)+((-C_A+2\,C_F)\,(d^4\,(40542624-190243274\,t+192068322\,t^2+3836748\,t^3-66949564$

$\times\,t^4+26848318\,t^5-4728326\,t^6)+d^6\,(1882471-8913958\,t+8102789\,t^2+356460\,t^3-2207455\,t^4+1127770\,t^5-267437\,t^6)+d^8$

$\times\,(13895-61460\,t+37227\,t^2+10624\,t^3-3427\,t^4+5844\,t^5-2319\,t^6)+8\,d^9\,(-51+210\,t-65\,t^2-68\,t^3-21\,t^4-14\,t^5+9\,t^6)-256$

$\times\,(-232830+946428\,t-926439\,t^2-93946\,t^3+431124\,t^4-144572\,t^5+17715\,t^6)+d^7\,(-211589+979788\,t-777821\,t^2-83840$

$\times\,t^3+163777\,t^4-111628\,t^5+32801\,t^6)+32\,d\,(-4606020+19626520\,t-19692823\,t^2-1407574\,t^3+8584510\,t^4-2967200$

$\times\,t^5+391475\,t^6)+d^5\,(-10734969+50985136\,t-49736701\,t^2-1089072\,t^3+15651861\,t^4-6887936\,t^5+1385313\,t^6)-8\,d^2$

$\times\,(-20082310+89080207\,t-90711499\,t^2-4305728\,t^3+36950986\,t^4-13253371\,t^5+1900051\,t^6)+4\,d^3\,(-25290911+115906328$

$\times\,t-118418999\,t^2-3523762\,t^3+44843377\,t^4-16865734\,t^5+2658741\,t^6)))/((-3+d)^2\,(1120-1244\,d+509\,d^2-91\,d^3+6$

$\times\,d^4)\,(-1+t)^4\,(6-10\,t+d\,(-1+3\,t))\,(24+8\,t-4\,t^2+d\,(-7-2\,t+t^2))))\,master[MI101MI101])/16-(C_F^2\,(-5+d)\,(-2+d)$

$\times\,(16-7\,d+d^2)\,N^2\,master[MI101MI201])/((12-7\,d+d^2)\,(-1+t))+(C_F\,N^2\,((-128\,(126-177\,d+87\,d^2-17\,d^3+d^4)\,I_2(R)$

$\times\,t)/((-40+63\,d-26\,d^2+3\,d^3)\,(-1+t)^3)-(4\,C_F\,(2-3\,d+d^2)\,(-116-24\,d^2+2\,d^3-d\,(-93+t)+8\,t))/((12-7\,d+d^2)$

$\times\,(-1+t)^2)-(8\,C_A\,((-5+d)^2\,(-2+d)+(362-181\,d+7\,d^2+4\,d^3)\,t+(-442+719\,d-320\,d^2+43\,d^3)\,t^2+(-662+689\,d-227\,d^2+24$

$\times\,d^3)\,t^3))/((15-8\,d+d^2)\,(1-t)^3\,(1+t))+(8\,(C_A-2\,C_F)\,(2\,d^5\,t-2\,d^4\,t\,(19+3\,t)-16\,(5-6\,t+45\,t^2)+d^3\,(3+230\,t+85\,t^2)-d^2$

$\times\,(29+554\,t+433\,t^2)+d\,(86+424\,t+934\,t^2)))/((-96+92\,d-29\,d^2+3\,d^3)\,(-1+t)^3)+(4\,C_F\,(7716+4732\,t-30228\,t^2-2284$

$\times\,t^3+d^5\,(-4+50\,t+92\,t^2+6\,t^3)-d^4\,(-101+633\,t+1477\,t^2+103\,t^3)+d^3\,(-949+2843\,t+9481\,t^2+697\,t^3)-2\,d^2\,(-2123+2427$

$\times\,t+15175\,t^2+1153\,t^3)+d\,(-9166+462\,t+48234\,t^2+3694\,t^3)))/((-60+47\,d-12\,d^2+d^3)\,(-1+t)^3\,(1+t))+((C_A-2\,C_F)\,(8$

$\times\,d^8\,t\,(95+475\,t-322\,t^2-298\,t^3+35\,t^4+15\,t^5)-2\,d^7\,(168+6767\,t+52572\,t^2-31708\,t^3-33170\,t^4+3549\,t^5+1726\,t^6)+128$

$\times\,(12600-199170\,t+459634\,t^2+713\,t^3-298323\,t^4+3787\,t^5+19499\,t^6)+d^6\,(8264+71561\,t+1259839\,t^2-663370\,t^3-799822$

$\times\,t^4+76609\,t^5+43047\,t^6)-2\,d^5\,(42908-71157\,t+4270363\,t^2-1911046\,t^3-2727906\,t^4+228675\,t^5+151875\,t^6)-16\,d$

$\times\,(220320-3109828\,t+9123624\,t^2-886421\,t^3-5923137\,t^4+169287\,t^5+375639\,t^6)+d^4\,(487656-3312979\,t+35834935$

$\times\,t^2-13098066\,t^3-23023118\,t^4+1637909\,t^5+1325631\,t^6)+4\,d^2\,(812228-9950382\,t+39203701\,t^2-7480656\,t^3-25402494$

$\times\,t^4+1106418\,t^5+1561385\,t^6)-2\,d^3\,(818780-8160169\,t+47645748\,t^2-13345304\,t^3-30761130\,t^4+1775673\,t^5+1830722\,t^6)))/$

$$((-3360 + 4852\,d - 2771\,d^2 + 782\,d^3 - 109\,d^4 + 6\,d^5)\,(-1+t)^4\,(1+t)\,(24 + 8\,t - 4\,t^2 + d\,(-7 - 2\,t + t^2))))\,master[MI101MI202])/4$$
$$+ (2\,C_F^2\,(-5+d)\,(16 - 7\,d + d^2)\,N^2\,master[MI201MI202])/((-4+d)\,(-1+t))$$
$$- (2\,C_F^2\,(-5+d)\,N^2\,((-4+d)^2\,(-3+d) - (-40 + 34\,d - 9\,d^2 + d^3)\,t)\,master[MI202MI202])/((8 - 6\,d + d^2)\,(-1+t)^2) + (C_F$$
$$\times\,(-8+3\,d)\,N^2\,(C_A\,t\,(-144\,d^8\,(-1 + 24\,t + 32\,t^2 - 318\,t^3 + 89\,t^4 + 30\,t^5) + 3\,d^7\,(-851 + 24947\,t + 66458\,t^2 - 378242\,t^3 + 115833$$
$$\times\,t^4 + 28623\,t^5) + 384\,(3714 + 155249\,t - 864954\,t^2 + 1315176\,t^3 - 725415\,t^4 + 178600\,t^5) + 4\,d^4\,(-28835 + 91206\,t - 32346126$$
$$\times\,t^2 + 76660546\,t^3 - 30987715\,t^4 + 618932\,t^5) - d^6\,(-15253 + 619267\,t + 3164698\,t^2 - 12360526\,t^3 + 4150731\,t^4 + 644571$$
$$\times\,t^5) - 192\,d\,(7721 + 609851\,t - 3899817\,t^2 + 6275274\,t^3 - 3247577\,t^4 + 683650\,t^5) + 2\,d^5\,(-10939 + 1065912\,t + 13145512$$
$$\times\,t^2 - 38747598\,t^3 + 14281515\,t^4 + 928494\,t^5) - 8\,d^3\,(-50291 + 3341736\,t - 49100134\,t^2 + 98260106\,t^3 - 43422009\,t^4 + 4170172$$
$$\times\,t^5) + 16\,d^2\,(2322 + 5391223\,t - 45327004\,t^2 + 79843104\,t^3 - 38347256\,t^4 + 6143159\,t^5)) + 2\,C_F\,(d^5\,(272240 - 6011238$$
$$\times\,t + 35204948\,t^2 - 38152808\,t^3 - 51180652\,t^4 + 44379614\,t^5 - 5281224\,t^6) + 72\,d^8\,(-1 + 16\,t - 201\,t^2 + 408\,t^3 + 325\,t^4 - 280$$
$$\times\,t^5 + 21\,t^6) - 3\,d^7\,(-796 + 14141\,t - 135729\,t^2 + 227434\,t^3 + 206794\,t^4 - 180327\,t^5 + 16323\,t^6) + 384\,(-11025 + 373971$$
$$\times\,t - 946816\,t^2 + 89277\,t^3 + 1494736\,t^4 - 1118973\,t^5 + 168180\,t^6) + d^6\,(-34004 + 673815\,t - 5005133\,t^2 + 6822018$$
$$\times\,t^3 + 7368718\,t^4 - 6439113\,t^5 + 679203\,t^6) - 192\,d\,(-46305 + 1467987\,t - 4206547\,t^2 + 1012824\,t^3 + 6504854\,t^4 - 5068166$$
$$\times\,t^5 + 755043\,t^6) + 4\,d^4\,(-335936 + 8222935\,t - 38838322\,t^2 + 32361274\,t^3 + 56691866\,t^4 - 48346013\,t^5 + 6303388\,t^6) + 16\,d^2$$
$$\times\,(-507054 + 14884704\,t - 49221825\,t^2 + 20286854\,t^3 + 74456768\,t^4 - 60112220\,t^5 + 8738281\,t^6) - 8\,d^3\,(-524562 + 14124685$$
$$\times\,t - 55141406\,t^2 + 33610736\,t^3 + 81707298\,t^4 - 68007891\,t^5 + 9467192\,t^6)))\,master[MI301p])/(48\,(-2+d)\,(-7+2\,d)\,(-14+3$$
$$\times\,d)\,(-10 + 3\,d)\,(12 - 7\,d + d^2)\,(-1+t)^4\,t\,(1+t)\,(6 - 10\,t + d\,(-1 + 3\,t))) + (C_F\,N^2\,(2\,C_F\,(-1+d)\,(8\,d^8\,(-63 + 1297\,t + 2062$$
$$\times\,t^2 + 134\,t^3 - 319\,t^4 + 25\,t^5) - d^7\,(-15084 + 214955\,t + 400017\,t^2 + 27329\,t^3 - 69101\,t^4 + 6444\,t^5) + 256\,(-52920 - 207696$$
$$\times\,t + 543414\,t^2 + 147806\,t^3 - 174185\,t^4 + 24092\,t^5) + 2\,d^6\,(-96830 + 923673\,t + 2139496\,t^2 + 165257\,t^3 - 411390\,t^4 + 43606$$
$$\times\,t^5) - 32\,d\,(-973728 - 2580416\,t + 11136896\,t^2 + 2481126\,t^3 - 3332665\,t^4 + 457272\,t^5) - d^5\,(-1400176 + 8281675\,t + 26361489$$
$$\times\,t^2 + 2440765\,t^3 - 5613873\,t^4 + 653600\,t^5) + 2\,d^4\,(-3128264 + 9560510\,t + 51098629\,t^2 + 5867226\,t^3 - 11974865\,t^4 + 1491688$$
$$\times\,t^5) + 8\,d^2\,(-3898884 - 4332110\,t + 49822383\,t^2 + 8973238\,t^3 - 13837943\,t^4 + 1864160\,t^5) - 4\,d^3\,(-4433532 + 3367397$$
$$\times\,t + 63723119\,t^2 + 9178863\,t^3 - 16316393\,t^4 + 2130688\,t^5)) + t\,(128\,(-2+d)^2\,(-924 + 1391\,d - 826\,d^2 + 242\,d^3 - 35\,d^4 + 2$$
$$\times\,d^5)\,I_2(R)\,(1 + 2\,n_f)\,(-1+t)^2\,(24 + 8\,t - 4\,t^2 + d\,(-7 - 2\,t + t^2)) + C_A\,(d^6\,(4549201 - 21709319\,t - 5691497\,t^2 + 2959247$$
$$\times\,t^3 - 137104\,t^4) + d^7\,(-413421 + 3744479\,t + 1155677\,t^2 - 453039\,t^3 - 9016\,t^4) - 16\,d^9\,(-39 - 1024\,t - 366\,t^2 + 104\,t^3 + 13$$
$$\times\,t^4) + d^8\,(10527 - 373607\,t - 126727\,t^2 + 40991\,t^3 + 3200\,t^4) - 256\,(-181896 + 251646\,t - 63134\,t^2 - 71095\,t^3 + 22696\,t^4) + 32$$
$$\times\,d\,(-5079088 + 7197580\,t - 1296202\,t^2 - 1900303\,t^3 + 574808\,t^4) + d^5\,(-26076815 + 80592997\,t + 15869183\,t^2 - 12607981$$
$$\times\,t^3 + 1450536\,t^4) - 8\,d^2\,(-29455526 + 45171773\,t - 4002218\,t^2 - 10786741\,t^3 + 2965976\,t^4) - 2\,d^4\,(-44488448 + 99932357$$
$$\times\,t + 11243808\,t^2 - 18129541\,t^3 + 3254776\,t^4) + 4\,d^3\,(-46701845 + 83300295\,t + 1140985\,t^2 - 17484697\,t^3 + 4090168\,t^4))))$$
$$\times\,master[MI301pt1])/(32\,(7 - 2\,d)^2\,(-2+d)\,(-1+d)\,(12 - 7\,d + d^2)\,(-1+t)^3\,t\,(24 + 8\,t - 4\,t^2 + d\,(-7 - 2\,t + t^2))) + (C_F\,N^2$$
$$\times\,(-(C_A\,t\,(d^8\,(3129 - 44658\,t - 626277\,t^2 + 235476\,t^3 + 4468911\,t^4 + 78750\,t^5 - 97155\,t^6) + 144\,d^9\,(-1 + 17\,t + 140\,t^2 - 166$$
$$\times\,t^3 - 1129\,t^4 - 43\,t^5 + 30\,t^6) + 768\,(7428 + 541396\,t - 308495\,t^2 - 4831901\,t^3 + 6562507\,t^4 - 1564505\,t^5 + 466750\,t^6) + d^7$$
$$\times\,(-25465 + 150480\,t + 8276003\,t^2 + 2720568\,t^3 - 54720067\,t^4 + 372792\,t^5 + 822105\,t^6) - 2\,d^6\,(-41445 - 1341429\,t + 30624678$$
$$\times\,t^2 + 31165346\,t^3 - 196406813\,t^4 + 6926835\,t^5 + 1224828\,t^6) - 384\,d\,(19156 + 2613124\,t - 2599668\,t^2 - 23868206\,t^3 + 35652691$$
$$\times\,t^4 - 7417044\,t^5 + 2106335\,t^6) - 4\,d^5\,(-6957 + 8520215\,t - 69919210\,t^2 - 123955048\,t^3 + 456809337\,t^4 - 30777743$$
$$\times\,t^5 + 2112926\,t^6) + 64\,d^2\,(25485 + 15942574\,t - 26098913\,t^2 - 153614457\,t^3 + 261701205\,t^4 - 45612643\,t^5 + 11894033\,t^6)$$

$$
\begin{aligned}
&+ 8\,d^4\left(-107961 + 23094155\,t - 101788596\,t^2 - 274652710\,t^3 + 715985417\,t^4 - 73203485\,t^5 + 12381932\,t^6\right) - 16\,d^3\left(-98260\right.\\
&+ 35444081\,t - 93833275\,t^2 - 371763980\,t^3 + 758001762\,t^4 - 104951633\,t^5 + 23639809\,t^6\left.\right)\big)\big) + 2\,C_F\left(72\,d^9\left(-1 + 15\,t - 161\,t^2 - 321\,t^3\right.\right.\\
&+ 1645\,t^4 + 1437\,t^5 - 331\,t^6 + 21\,t^7\left.\right) - 3\,d^8\left(-892 + 13201\,t - 103246\,t^2 - 295613\,t^3 + 1043072\,t^4 + 979543\,t^5 - 276150\,t^6 + 16341\,t^7\right)\\
&- 768\left(-22050 + 371622\,t + 984870\,t^2 - 4648723\,t^3 + 1086341\,t^4 + 7041741\,t^5 - 4349711\,t^6 + 226810\,t^7\right) + d^7\left(-43556 + 647143\,t\right.\\
&- 3473372\,t^2 - 14132645\,t^3 + 36437692\,t^4 + 37503853\,t^5 - 12705660\,t^6 + 709137\,t^7\left.\right) + 384\,d\left(-103635 + 1745228\,t + 3518856\,t^2\right.\\
&- 23134102\,t^3 + 9264323\,t^4 + 35682873\,t^5 - 21820914\,t^6 + 1117831\,t^7\left.\right) - 2\,d^6\left(-204128 + 3083469\,t - 10372501\,t^2 - 63113640\,t^3\right.\\
&+ 122460946\,t^4 + 141508089\,t^5 - 56374909\,t^6 + 3006210\,t^7\left.\right) - 64\,d^2\left(-645969 + 10748370\,t + 14196732\,t^2 - 153407374\,t^3 + 91692708\,t^4\right.\\
&+ 244208196\,t^5 - 144345033\,t^6 + 7303486\,t^7\left.\right) + 4\,d^5\left(-608176 + 9410833\,t - 16604331\,t^2 - 176756914\,t^3 + 260959440\,t^4 + 348641091\,t^5\right.\\
&- 1594181197\,t^6 + 8233630\,t^7\left.\right) - 8\,d^4\left(-1196434 + 19017661\,t - 9511891\,t^2 - 324997004\,t^3 + 363772512\,t^4 + 582372831\,t^5\right.\\
&- 297840643\,t^6 + 15100072\,t^7\left.\right) + 16\,d^3\left(-1556178 + 25372808\,t + 12788919\,t^2 - 394786933\,t^3 + 328951818\,t^4 + 659734946\,t^5\right.\\
&- 367512947\,t^6 + 18516359\,t^7\left.\right)\big)\big)\,master[MI302p]\big)/\big(48\,(-3 + d)^2\left(-7840 + 12152\,d - 7340\,d^2 + 2170\,d^3 - 315\,d^4 + 18\,d^5\right)(-1 + t)^4\,t\\
&\times (1 + t)(6 - 10\,t + d(-1 + 3\,t))\big) + \big(C_F\,N^2\big((4\,I_2(R)\,(d(98 + 344\,t - 82\,t^2) + d^3(30 + 72\,t - 14\,t^2) + d^4(-3 - 6\,t + t^2) + 8(-7 - 13\,t + 2\,t^2)\\
&+ d^2(-89 - 266\,t + 59\,t^2)))/(-24 + 34\,d - 11\,d^2 + d^3) + ((C_A - 2\,C_F)(d^5(51 + 249\,t + 117\,t^2 - 9\,t^3) + d^6(-3 - 9\,t - 5\,t^2 + t^3) - d^4(305\\
&+ 2367\,t + 1079\,t^2 + 25\,t^3) - 32(-121 + 393\,t + 249\,t^2 + 95\,t^3) - 8\,d^2(-57 + 3160\,t + 1589\,t^2 + 308\,t^3) + d^3(653 + 10759\,t + 5059\,t^2 + 561\,t^3)\\
&+ 4\,d(-926 + 7281\,t + 4032\,t^2 + 1129\,t^3))))/((-4 + d)^2(6 - 5\,d + d^2)(1 + t)))\,master[MI303p]\big)/\big(2\,(-1 + t)^3\big) + \big(C_F\,N^2\,(C_A\,(-72\,d^7\,t\,(1\\
&- 6\,t + t^2) + d^3(43968 - 404848\,t + 698252\,t^2 - 228634\,t^3) + 3\,d^6(-32 + 716\,t - 2565\,t^2 + 551\,t^3) + 64(-672 + 3808\,t - 5010\,t^2 + 1745\,t^3)\\
&- 2\,d^5(-824 + 11734\,t - 29946\,t^2 + 7801\,t^3) - 8\,d(-12484 + 82596\,t - 116870\,t^2 + 41019\,t^3) + d^4(-11696 + 130048\,t - 263189\,t^2\\
&+ 78723\,t^3) + 4\,d^2(-22972 + 178612\,t - 275203\,t^2 + 94801\,t^3)) + 2(-32(-2 + d)^2(308 - 361\,d + 155\,d^2 - 29\,d^3 + 2\,d^4)\,I_2(R)(1 + 2\,n_f)\\
&\times (-1 + t)^2(1 + t) + C_F(-1 + d)(d^4(984 + 6700\,t + 16659\,t^2 - 2791\,t^3) + 4\,d^6(1 + 35\,t + 35\,t^2 + t^3) + d^5(-94 - 1682\,t - 2243\,t^2 + 137\,t^3)\\
&- 64(-308 + 1288\,t - 2284\,t^2 + 845\,t^3) + 2\,d^3(-2808 - 2032\,t - 36044\,t^2 + 9333\,t^3) + 8\,d(-3686 + 13458\,t - 31918\,t^2 + 11313\,t^3)\\
&- 4\,d^2(-4462 + 10194\,t - 46105\,t^2 + 14765\,t^3))))\,master[MI402p]\big)/\big(4\,(-4 + d)(-2 + d)(-1 + d)(-7 + 2\,d)(-8 + 3\,d)(-1 + t)^3(1 + t)\big)\\
&+ \big((C_A - 2\,C_F)\,C_F\,N^2\,(d^5(809 + 4062\,t - 13112\,t^2 + 5210\,t^3 - 2185\,t^4) + d^6(-37 - 208\,t + 694\,t^2 - 240\,t^3 + 111\,t^4) + 64(-654 - 3787\,t\\
&+ 10835\,t^2 - 6615\,t^3 + 1565\,t^4) + 2\,d^4(-3539 - 16285\,t + 51757\,t^2 - 23211\,t^3 + 8782\,t^4) - 16\,d(-5608 - 27317\,t + 82133\,t^2 - 47753\,t^3\\
&+ 12793\,t^4) - 4\,d^3(-7887 - 34693\,t + 109163\,t^2 - 54317\,t^3 + 18430\,t^4) + 8\,d^2(-9383 - 41845\,t + 129641\,t^2 - 70359\,t^3 + 21274\,t^4))\\
&\times master[MI403p]\big)/\big(8\,(-4 + d)(-3 + d)(14 - 11\,d + 2\,d^2)(-1 + t)^3(6 - 10\,t + d(-1 + 3\,t))\big) - \big((C_A - 2\,C_F)\,C_F\,(-5 + d)\,N^2\,t^2\,(d\,(3012\\
&+ 284\,t - 6228\,t^2) + d^3(439 - 168\,t - 531\,t^2) + d^4(37\,t^2 + 16\,t - 37) + 8\,d^2(-225 + 54\,t + 347\,t^2) + 8(626\,t^2 - 187\,t - 218))\,master[MI404p]\big)\\
&/\big((d - 2)(2\,d - 7)(12 - 7\,d + d^2)(-1 + t)^3(6 - 10\,t + d(-1 + 3\,t))\big) + \big(2\,C_F\,N^2\,((4\,I_2(R)\,(40(1 + t + t^2) + d^4(1 + 6\,t + t^2) - 2\,d^3(3 + 34\,t + 3\,t^2)\\
&- 2\,d(7 + 138\,t + 7\,t^2) + d^2(9 + 238\,t + 9\,t^2)))/(3\,d^3 - 26\,d^2 + 63\,d - 40) + ((C_A - 2\,C_F)(d^6(1 + 6\,t + t^2) - d^5(11 + 167\,t + 17\,t^2 + 9\,t^3) + 64(50\,t^3\\
&- 33\,t^2 + 210\,t - 73) + d^4(-12 + 1653\,t + 94\,t^2 + 153\,t^3) - 2\,d^3(-305 + 3994\,t + 61\,t^2 + 508\,t^3) - 8\,d(-782 + 3289\,t - 278\,t^2 + 650\,t^3) + 4\,d^2\\
&\times (-768 + 5097\,t - 152\,t^2 + 823\,t^3)))/(192 - 280\,d + 150\,d^2 - 35\,d^3 + 3\,d^4))\,master[MI407p]\big)/(-1 + t)^3 + \big(4\,C_F\,N^2\,((4(-26 + 23\,d - 8\,d^2 + d^3)\\
&\times I_2(R)(1 + t)^2)/(-40 + 63\,d - 26\,d^2 + 3\,d^3) + ((C_A - 2\,C_F)(d^5(1 + t)^2 - d^4(13 + 53\,t + 19\,t^2 + 3\,t^3) - 64(-4 + 15\,t + 6\,t^2 + 5\,t^3) + d^3(54 + 405\,t\\
&+ 136\,t^2 + 41\,t^3) - 2\,d^2(26 + 661\,t + 228\,t^2 + 101\,t^3) + 2\,d(212\,t^3 + 351\,t^2 + 954\,t - 73)))/(192 - 280\,d + 150\,d^2 - 35\,d^3 + 3\,d^4))\,master[MI408p]\big)\\
&/(-1 + t^2) + \big(C_F\,(-C_A + 2\,C_F)\,N^2\,(d^3(89360 - 379664\,t + 46678\,t^2 + 65588\,t^3 - 6106\,t^4) + d^5(218 - 8561\,t + 1449\,t^2 + 1617\,t^3 - 163\,t^4) + 8\,d^6\\
&\times (4 + 45\,t - 9\,t^2 - 9\,t^3 + t^4) + 64(-5986 + 11048\,t - 821\,t^2 - 1654\,t^3 + 157\,t^4) + d^4(-10566 + 79833\,t - 11503\,t^2 - 14405\,t^3 + 1377\,t^4) - 8\,d\\
&\times (-71652 + 163408\,t - 14523\,t^2 - 25726\,t^3 + 2393\,t^4) + 4\,d^2(-82494 + 244921\,t - 25673\,t^2 - 40445\,t^3 + 3735\,t^4))\,master[MI409p]\big)\\
&/\big(2\,(6\,d^3 - 49\,d^2 + 130\,d - 112)(-1 + t)^3(24 + 8\,t - 4\,t^2 + d(-7 - 2\,t + t^2))\big)
\end{aligned}
$$

$$
\begin{aligned}
&+ ((C_A - 2\,C_F)\,C_F\,N^2\,(24\,d^6\,(2 - t - 2\,t^2 + t^3) + d^2\,(77952 + 21408\,t - 184160\,t^2 + 86384\,t^3 - 4700\,t^4) + d^4\,(9498 - 3275 \\
&\quad \times t - 12775\,t^2 + 6501\,t^3 - 241\,t^4) + d^5\,(-1114 + 515\,t + 1255\,t^2 - 647\,t^3 + 15\,t^4) - 64\,(-260 - 1166\,t + 2378\,t^2 - 997 \\
&\quad \times t^3 + 66\,t^4) + 2\,d^3\,(-19318 + 2461\,t + 33011\,t^2 - 16226\,t^3 + 760\,t^4) + 8\,d\,(-8696 - 10100\,t + 32986\,t^2 - 14661\,t^3 + 890\,t^4)) \\
&\quad \times master[MI410p])/((448 - 632\,d + 326\,d^2 - 73\,d^3 + 6\,d^4)\,(-1 + t)^2\,(24 + 8\,t - 4\,t^2 + d\,(-7 - 2\,t + t^2))) + ((C_A - 2\,C_F)\,C_F\,N^2\,(16 \\
&\quad \times d^6\,(-1 - 2\,t + t^2) + d^5\,(30 + 951\,t - 342\,t^2 - 15\,t^3) + 64\,(1530 - 2392\,t + 407\,t^2 + 66\,t^3) - 8\,d^3\,(2606 - 7257\,t + 1726\,t^2 + 190 \\
&\quad \times t^3) + d^4\,(2194 - 10537\,t + 2994\,t^2 + 241\,t^3) - 8\,d\,(17860 - 31944\,t + 5939\,t^2 + 890\,t^3) + 4\,d^2\,(20010 - 42653\,t + 8841\,t^2 + 1175 \\
&\quad \times t^3))\,master[MI411p])/(2\,(336 - 502\,d + 277\,d^2 - 67\,d^3 + 6\,d^4)\,(-1 + t)\,(24 + 8\,t - 4\,t^2 + d\,(-7 - 2\,t + t^2))) - (C_F\,N^2\,(C_A\,t\,(d^3 \\
&\quad \times (-668 - 2093\,t + 31003\,t^2 + 961\,t^3 - 643\,t^4) + d^5\,(-4 + 80\,t + 424\,t^2 + 80\,t^3 - 4\,t^4) + d^4\,(86 - 455\,t - 5721\,t^2 - 713\,t^3 + 83\,t^4) + 16 \\
&\quad \times (196 + 2587\,t - 5094\,t^2 + 1676\,t^3 + 173\,t^4) - 4\,d\,(1106 + 13098\,t - 31791\,t^2 + 7206\,t^3 + 1033\,t^4) + d^2\,(2464 + 21058\,t - 86284 \\
&\quad \times t^2 + 7986\,t^3 + 2360\,t^4)) + 2\,C_F\,(2\,d^5\,(1 - 11\,t + 154\,t^2 + 154\,t^3 - 11\,t^4 + t^5) - d^4\,(43 - 581\,t + 3967\,t^2 + 4361\,t^3 - 538\,t^4 + 44\,t^5) - 16 \\
&\quad \times (147 - 2233\,t + 781\,t^2 + 5392\,t^3 - 1432\,t^4 + 117\,t^5) + d^3\,(358 - 5284\,t + 19561\,t^2 + 25693\,t^3 - 4547\,t^4 + 363\,t^5) + 4\,d\,(735 - 11309 \\
&\quad \times t + 11363\,t^2 + 32308\,t^3 - 8162\,t^4 + 661\,t^5) - 2\,d^2\,(730 - 11164\,t + 22519\,t^2 + 39796\,t^3 - 8863\,t^4 + 710\,t^5)))\,master[MI501p])/(2 \\
&\quad \times (-7 + 2\,d)\,(6 - 5\,d + d^2)\,(-1 + t)^3\,t\,(1 + t)) - (C_F\,(-9 + 2\,d)\,N^2\,(C_A\,t\,(-16\,d^4\,(1 + 10\,t + t^2) + d^3\,(-79 + 1608\,t + 7\,t^2) + 12 \\
&\quad \times (418 - 777\,t + 303\,t^2) - 6\,d\,(869 - 1975\,t + 570\,t^2) + d^2\,(1551 - 6290\,t + 851\,t^2)) + 2\,C_F\,(d\,(-2352 + 2654\,t + 9942\,t^2 - 2036\,t^3) + d^3 \\
&\quad \times (-140 + 915\,t + 1092\,t^2 - 139\,t^3) + 8\,d^4\,(1 - 13\,t - 13\,t^2 + t^3) + 12\,(196 + 2\,t - 679\,t^2 + 145\,t^3) + d^2\,(872 - 2687\,t - 4726\,t^2 + 829\,t^3))) \\
&\quad \times master[MI502p])/(2\,(-7 + 2\,d)\,(-24 + 26\,d - 9\,d^2 + d^3)\,(-1 + t)\,t\,(1 + t)) + (C_F\,(-C_A + 2\,C_F)\,(-4 + d)\,N^2\,(20 - 22\,t + d\,(-3 + 5 \\
&\quad \times t))\,master[MI601p])/(-28 + 8\,d)) + S3\,((C_F\,(-2 + d)\,N^2\,((-16\,I_2(R)\,(32\,(-7 + d)^2\,(-15 + 23\,d - 9\,d^2 + d^3) + 2\,(-2904 + 7178 \\
&\quad \times d - 4981\,d^2 + 1430\,d^3 - 179\,d^4 + 8\,d^5)\,t + (-672 + 1468\,d - 1120\,d^2 + 377\,d^3 - 56\,d^4 + 3\,d^5)\,t^2))/((35 - 12\,d + d^2)^2\,(8 - 11\,d + 3 \\
&\quad \times d^2)\,t) + (8\,C_F\,((-3 + d)^2\,(336 - 324\,d + 116\,d^2 - 18\,d^3 + d^4) - 2\,(18192 - 23626\,d + 12170\,d^2 - 3162\,d^3 + 436\,d^4 - 31\,d^5 + d^6) \\
&\quad \times t + (13152 - 19232\,d + 11096\,d^2 - 3246\,d^3 + 503\,d^4 - 38\,d^5 + d^6)\,t^2 - 8\,(-58 + 59\,d - 19\,d^2 + 2\,d^3)\,t^3))/((-5 + d)\,(12 - 7\,d + d^2)^2 \\
&\quad \times (-1 + t)^2\,t) + (16\,I_2(R)\,(-7551360 + 17497812\,d - 17855908\,d^2 + 10513945\,d^3 - 3929328\,d^4 + 965115\,d^5 - 155663\,d^6 + 15896 \\
&\quad \times d^7 - 933\,d^8 + 24\,d^9 + (-6544992 + 15524244\,d - 15477276\,d^2 + 8517125\,d^3 - 2839832\,d^4 + 589100\,d^5 - 74414\,d^6 + 5243\,d^7 - 158 \\
&\quad \times d^8)\,t + (12588288 - 29037636\,d + 28875196\,d^2 - 16269417\,d^3 + 5734162\,d^4 - 1314230\,d^5 + 196476\,d^6 - 18541\,d^7 + 1006\,d^8 - 24 \\
&\quad \times d^9)\,t^2 + (-1120800 + 2525820\,d - 2404292\,d^2 + 1255487\,d^3 - 392364\,d^4 + 75100\,d^5 - 8594\,d^6 + 537\,d^7 - 14\,d^8)\,t^3 + (-6 + d)^2 \\
&\quad \times (1344 - 3496\,d + 3482\,d^2 - 1725\,d^3 + 451\,d^4 - 59\,d^5 + 3\,d^6)\,t^4))/((35 - 12\,d + d^2)^2\,(-576 + 1224\,d - 914\,d^2 + 313\,d^3 - 50\,d^4 + 3\,d^5) \\
&\quad \times (1 - t)^3) + (16\,C_F\,(-2 + d)\,(-1 + d)\,(-594 + d^3\,(35 - 31\,t) + 2\,d^4\,(-1 + t) + 508\,t - 34\,t^2 + d^2\,(-223 + 185\,t - 2\,t^2) + 4\,d\,(152 - 125 \\
&\quad \times t + 4\,t^2)))/((-3 + d)^2\,(20 - 9\,d + d^2)\,(-1 + t)^2) - (4\,C_A\,(2196 + 1248\,t - 276\,t^2 + d^2\,(276 + 84\,t - 72\,t^2) + d^3\,(5 - 6\,t + t^2) + d^4\,(-3 + 2 \\
&\quad \times t + t^2) + 2\,d\,(-763 - 332\,t + 135\,t^2)))/((-3 + d)^2\,(20 - 9\,d + d^2)\,(-1 + t)^2) - (4\,(C_A - 2\,C_F)\,(12\,d^9\,t\,(-1 + t^2) + d^6\,(26933 - 49607 \\
&\quad \times t + 62098\,t^2 - 185454\,t^3 + 41673\,t^4 - 2843\,t^5) + 3\,d^8\,(17 + 47\,t + 88\,t^2 - 252\,t^3 + 39\,t^4 - 3\,t^5) + d^7\,(-1777 + 1875\,t - 6342 \\
&\quad \times t^2 + 16394\,t^3 - 3353\,t^4 + 243\,t^5) - 64\,(-81627 + 59920\,t + 71334\,t^2 + 192960\,t^3 - 38267\,t^4 + 1920\,t^5) + d^5\,(-231701 + 444809 \\
&\quad \times t - 308150\,t^2 + 1264858\,t^3 - 293253\,t^4 + 18861\,t^5) + 8\,d\,(-1285139 + 1256138\,t + 852556\,t^2 + 3480414\,t^3 - 729473\,t^4 + 38288 \\
&\quad \times t^5) - 2\,d^4\,(-617876 + 1085477\,t - 364243\,t^2 + 2756091\,t^3 - 638777\,t^4 + 38848\,t^5) - 8\,d^2\,(-1090713 + 1351984\,t + 402636 \\
&\quad \times t^2 + 3435174\,t^3 - 753795\,t^4 + 41434\,t^5) + 2\,d^3\,(-2088649 + 3151158\,t - 77130\,t^2 + 7765644\,t^3 - 1763901\,t^4 + 101806 \\
&\quad \times t^5)))/((-8 + 3\,d)\,(-60 + 47\,d - 12\,d^2 + d^3)^2\,(-1 + t)^2\,(1 + t)\,(6 - 10\,t + d\,(-1 + 3\,t))) + ((-C_A + 2\,C_F)\,(d^7\,(-1649241 + 1846124 \\
&\quad \times t + 1563459\,t^2 - 1521344\,t^3 - 1290275\,t^4 + 1378132\,t^5 - 307527\,t^6) + d^9\,(-13657 + 28556\,t - 14525\,t^2 - 4800\,t^3 + 1261\,t^4 + 4820 \\
&\quad \times t^5 - 1655\,t^6) + 3\,d^{10}\,(133 - 326\,t + 259\,t^2 - 4\,t^3 - 53\,t^4 - 22\,t^5 + 13\,t^6) + 2048\,(-59454 + 236670\,t - 232795\,t^2 - 54802\,t^3
\end{aligned}
$$

$$+168652\,t^4 - 68846\,t^5 + 8055\,t^6) + d^8\,(200939 - 334698\,t + 27937\,t^2 + 132220\,t^3 + 56101\,t^4 - 113170\,t^5 + 29903\,t^6)$$
$$-256\,d\,(-906540 + 4519500\,t - 5164151\,t^2 - 676250\,t^3 + 3516566\,t^4 - 1539748\,t^5 + 194631\,t^6)$$
$$+4\,d^6\,(2032241 - 624932\,t - 5019156\,t^2 + 2407958\,t^3 + 3205967\,t^4 - 2557306\,t^5 + 502716\,t^6)$$
$$+64\,d^2\,(-2497120 + 18450734\,t - 24886805\,t^2 - 837690\,t^3 + 16074476\,t^4 - 7613466\,t^5 + 1041063\,t^6)$$
$$+16\,d^4\,(1929084 + 12922370\,t - 28925412\,t^2 + 4777565\,t^3 + 17447009\,t^4 - 10082915\,t^5 + 1627747\,t^6)$$
$$-4\,d^5\,(5867088 + 7345151\,t - 31050675\,t^2 + 8998950\,t^3 + 18724142\,t^4 - 12402349\,t^5 + 2195741\,t^6)$$
$$-16\,d^3\,(-1561021 + 41140713\,t - 68113122\,t^2 + 4250220\,t^3 + 42188221\,t^4 - 21877001\,t^5 + 3243638\,t^6)))/((12 - 7\,d + d^2)^2\,(-280$$
$$+241\,d - 67\,d^2 + 6\,d^3)\,(-1+t)^3\,(6 - 10\,t + d\,(-1 + 3\,t))\,(24 + 8\,t - 4\,t^2 + d\,(-7 - 2\,t + t^2))))\,master[MI101MI101])/64 + (C_F\,(-2$$
$$+d)\,N^2\,(-(((C_F\,(16 - 7\,d + d^2)\,(-3 + d + t))/((-4 + d)^2\,(-1 + t))) - (4\,(-3 + d)^2\,I_2(R)\,(6\,d^4 - 3\,d^3\,(34 + t) + 48\,(39 + t) + d^2$$
$$\times(668 + 29\,t) - 2\,d\,(946 + 37\,t)))/((-1120 + 2204\,d - 1461\,d^2 + 433\,d^3 - 59\,d^4 + 3\,d^5)\,t))\,master[MI101MI201])/(-3 + d)$$
$$+(C_F\,N^2\,((16\,(C_A - 2\,C_F)\,(-1 + t)\,((-5 + d)^2\,(16 - 14\,d + 3\,d^2) + (-5008 + 6618\,d - 3405\,d^2 + 855\,d^3 - 105\,d^4 + 5\,d^5)$$
$$\times\,t + (1088 - 1704\,d + 1054\,d^2 - 325\,d^3 + 50\,d^4 - 3\,d^5)\,t^2))/(480 - 556\,d + 237\,d^2 - 44\,d^3 + 3\,d^4) + (16\,C_A\,(-1+t)\,((-5+d)^2\,(8 - 6$$
$$\times\,d + d^2) + (-2326 + 2311\,d - 832\,d^2 + 128\,d^3 - 7\,d^4)\,t + (-1688 + 1434\,d - 389\,d^2 + 30\,d^3 + d^4)\,t^2 + (646 - 803\,d + 360\,d^2 - 70$$
$$\times\,d^3 + 5\,d^4)\,t^3))/((-60 + 47\,d - 12\,d^2 + d^3)\,(1 + t)) - (8\,C_F\,(2 - 3\,d + d^2)\,(-1 + t)\,(2\,d^3\,(-1 + t) + d\,(-93 + 80\,t - 7\,t^2) + d^2\,(24 - 21$$
$$\times\,t + t^2) + 4\,(29 - 27\,t + 4\,t^2)))/(12 - 7\,d + d^2) - (64\,(6 - 5\,d + d^2)\,I_2(R)\,t\,(6\,d^4\,(-1 + t) - 3\,d^3\,(-37 + 32\,t + t^2) + 16\,(-130 + 111$$
$$\times\,t + 3\,t^2) + d^2\,(-743 + 610\,t + 29\,t^2) - 2\,d\,(-1053 + 872\,t + 37\,t^2)))/(-1120 + 2204\,d - 1461\,d^2 + 433\,d^3 - 59\,d^4 + 3\,d^5) + (8\,C_F$$
$$\times\,(-1+t)\,(d^6\,(-5 - 8\,t + 13\,t^2) + d\,(46332 + 118276\,t - 46556\,t^2 - 4260\,t^3) + d^3\,(8734 + 22220\,t - 12162\,t^2 - 640\,t^3) + d^5\,(137 + 292$$
$$\times\,t - 281\,t^2 - 4\,t^3) + 2\,d^4\,(-758 - 1829\,t + 1266\,t^2 + 41\,t^3) + 16\,(-1989 - 4941\,t + 1693\,t^2 + 181\,t^3) + d^2\,(-27762 - 71618\,t + 32742$$
$$\times\,t^2 + 2390\,t^3)))/((-4 + d)^2\,(15 - 8\,d + d^2)\,(1 + t)) + ((C_A - 2\,C_F)\,(6\,d^9\,t\,(-175 + 300\,t - 26\,t^2 - 132\,t^3 + 33\,t^4) - d^8\,(672 - 43953$$
$$\times\,t + 61319\,t^2 + 2690\,t^3 - 26606\,t^4 + 5791\,t^5 + 87\,t^6) + 2\,d^7\,(9608 - 384363\,t + 456616\,t^2 + 76738\,t^3 - 196452\,t^4 + 36321\,t^5 + 1340$$
$$\times\,t^6) - 1024\,(12600 - 292662\,t + 171640\,t^2 + 181705\,t^3 - 71971\,t^4 - 5089\,t^5 + 2517\,t^6) - d^6\,(237744 - 7521653\,t + 7818179$$
$$\times\,t^2 + 2261746\,t^3 - 3347646\,t^4 + 507283\,t^5 + 35451\,t^6) + 128\,d\,(245520 - 5678932\,t + 3615192\,t^2 + 3321421\,t^3 - 1528255$$
$$\times\,t^4 - 45111\,t^5 + 47209\,t^6) + 2\,d^5\,(830920 - 22935691\,t + 21227911\,t^2 + 8703266\,t^3 - 9066774\,t^4 + 1065113\,t^5 + 131639\,t^6) - 32$$
$$\times\,d^2\,(1032548 - 24147250\,t + 16737170\,t^2 + 13141015\,t^3 - 7109131\,t^4 + 66789\,t^5 + 190367\,t^6) - 4\,d^4\,(1794092 - 45501895$$
$$\times\,t + 37951376\,t^2 + 20188348\,t^3 - 16186026\,t^4 + 1336395\,t^5 + 300350\,t^6) + 8\,d^3\,(2449788 - 58977045\,t + 44705604\,t^2 + 29353566$$
$$\times\,t^3 - 19038076\,t^4 + 887651\,t^5 + 431320\,t^6)))/((-4 + d)^2\,(840 - 1003\,d + 442\,d^2 - 85\,d^3 + 6\,d^4)\,(1 + t)\,(24 + 8\,t - 4\,t^2 + d\,(-7 - 2$$
$$\times\,t + t^2))))\,master[MI101MI202])/(16\,(-1 + t)^3) + (C_F^2\,(16 - 7\,d + d^2)^2\,N^2\,master[MI201MI201])/(-4 + d)^2 + (C_F^2\,(16 - 7$$
$$\times\,d + d^2)\,N^2\,(d\,(9 - 7\,t) + d^2\,(-1 + t) + 4\,(-5 + 4\,t))\,master[MI201MI202])/((-4 + d)^2\,(-1 + t)) + (C_F^2\,N^2\,((-4 + d)^3\,(15 - 8$$
$$\times\,d + d^2) - 2\,(-720 + 944\,d - 495\,d^2 + 132\,d^3 - 18\,d^4 + d^5)\,t + (-2 + d)\,(16 - 7\,d + d^2)^2\,t^2)\,master[MI202MI202])/((-4 + d)^2$$
$$\times\,(-2 + d)\,(-1 + t)^2) + (C_F\,(-8 + 3\,d)\,(C_A\,(-120960 + 189808\,d - 122112\,d^2 + 41026\,d^3 - 7525\,d^4 + 702\,d^5 - 25\,d^6) + 2\,C_F$$
$$\times\,(66080 - 110064\,d + 72896\,d^2 - 24466\,d^3 + 4309\,d^4 - 358\,d^5 + 9\,d^6))\,N^2\,master[MI301])/(8\,(-4 + d)^3\,(-3 + d)\,(-7 + 2\,d)\,t) + (C_F$$
$$\times\,(-8 + 3\,d)\,N^2\,(C_A\,t\,(9\,d^{10}\,(19 + 231\,t - 1646\,t^2 + 1770\,t^3 - 389\,t^4 + 15\,t^5) - 18\,d^9\,(275 + 6115\,t - 35190\,t^2 + 32350\,t^3 - 6893$$
$$\times\,t^4 + 1039\,t^5) + 15360\,(438 + 235280\,t - 867950\,t^2 + 697818\,t^3 - 269575\,t^4 + 76855\,t^5) + d^8\,(57257 + 2395389\,t - 11873698$$
$$\times\,t^2 + 9685902\,t^3 - 2064871\,t^4 + 525909\,t^5) - 2\,d^7\,(158867 + 14553397\,t - 64544438\,t^2 + 48322842\,t^3 - 10648989\,t^4 + 3574673$$
$$\times\,t^5) - 768\,d\,(-14452 + 13173541\,t - 46802640\,t^2 + 35435297\,t^3 - 12744785\,t^4 + 4055685\,t^5) + 4\,d^6\,(155832 + 55500561$$
$$\times\,t - 226085618\,t^2 + 160162734\,t^3 - 37518766\,t^4 + 14461113\,t^5) + 128\,d^2\,(-327045 + 95983697\,t - 337053911\,t^2 + 243185148$$
$$\times\,t^3 - 80733973\,t^4 + 28312114\,t^5) - 8\,d^5\,(-274170 + 139930255\,t - 534496616\,t^2 + 368198128\,t^3 - 93460066\,t^4 + 37813165\,t^5)$$

$$+16\,d^4\,(-1006043 + 237211636\,t - 865259198\,t^2 + 593645228\,t^3 - 164846201\,t^4 + 65912910\,t^5)$$
$$-32\,d^3\,(-1226525 + 267468306\,t - 948330794\,t^2 + 661470808\,t^3 - 201243959\,t^4 + 76376648\,t^5))$$
$$+2\,C_F\,(d^7\,(1252072 - 32733394\,t + 58959154\,t^2 + 65336980\,t^3 - 140206212\,t^4 + 66178494\,t^5 - 5263574\,t^6)$$
$$+9\,d^{10}\,(-16 + 349\,t - 855\,t^2 - 402\,t^3 + 1798\,t^4 - 1003\,t^5 + 129\,t^6) - 6\,d^9\,(-1012 + 23451\,t - 51133\,t^2 - 38478\,t^3$$
$$+112994\,t^4 - 59229\,t^5 + 6495\,t^6) + 15360\,(-11025 + 405345\,t - 534754\,t^2 - 631233\,t^3 + 1260586\,t^4 - 487125\,t^5 + 13620\,t^6)$$
$$+d^8\,(-113872 + 2804015\,t - 5517941\,t^2 - 5305614\,t^3 + 12709074\,t^4 - 6299969\,t^5 + 588435\,t^6)$$
$$-256\,d\,(-1686825 + 59692227\,t - 80545696\,t^2 - 98219151\,t^3 + 193476007\,t^4 - 76232745\,t^5 + 2431245\,t^6)$$
$$+128\,d^2\,(-3851655 + 130708835\,t - 181171811\,t^2 - 226789284\,t^3 + 441785712\,t^4 - 177756842\,t^5 + 6520031\,t^6)$$
$$+4\,d^6\,(-2236992 + 61978690\,t - 103581175\,t^2 - 124609818\,t^3 + 251834806\,t^4 - 113942944\,t^5 + 7721385\,t^6)$$
$$-8\,d^5\,(-5433748 + 159149438\,t - 249937899\,t^2 - 313939116\,t^3 + 615409304\,t^4 - 268530254\,t^5 + 15526907\,t^6)$$
$$-32\,d^3\,(-10364766 + 336068655\,t - 481455424\,t^2 - 612796212\,t^3 + 1185828744\,t^4 - 488408469\,t^5 + 20747596\,t^6)$$
$$+16\,d^4\,(-9094526 + 280737465\,t - 419038714\,t^2 - 534937532\,t^3 + 1035869472\,t^4 - 438193507\,t^5 + 21674994\,t^6)))\,master[MI301p])$$
$$/(192\,(-5+d)\,(-4+d)^2\,(-3+d)\,(-2+d)\,(-7+2\,d)\,(-14+3\,d)\,(-10+3\,d)\,(-1+t)^3\,t\,(1+t)\,(6 - 10\,t + d\,(-1+3\,t)))$$
$$+(C_F\,N^2\,((16\,(C_A - 2\,C_F)\,(13312 - 21024\,d + 14014\,d^2 - 5085\,d^3 + 1064\,d^4 - 122\,d^5 + 6\,d^6)\,(-1+t))/((-4+d)^2\,(21 - 13\,d + 2$$
$$\times d^2)) + (4\,C_A\,(-8+3\,d)\,(10+d\,(-3+t) - 6\,t)\,(-1+t))/(-28+43\,d - 17\,d^2 + 2\,d^3) + (4\,C_A\,(-8+3\,d)\,(-5+6\,d)\,(10+d\,(-3+t) - 6$$
$$\times t)\,(-1+t))/(-28+43\,d - 17\,d^2 + 2\,d^3) - (32\,(-2+d)\,(-8+3\,d)\,I_2(R)\,(10+d\,(-3+t) - 6\,t)\,(-1+t))/(-28+43\,d - 17\,d^2 + 2$$
$$\times d^3) - (64\,(-2+d)\,(-8+3\,d)\,I_2(R)\,n_f\,(10+d\,(-3+t) - 6\,t)\,(-1+t))/(-28+43\,d - 17\,d^2 + 2\,d^3) - (16\,C_A\,(-1+t)\,(-196\,d^4 + 12$$
$$\times d^5 + d^3\,(1258+3\,t) - 16\,(223+9\,t) - d^2\,(3947+35\,t) + 2\,d\,(3013+63\,t)))/((-4+d)^2\,(21 - 13\,d + 2\,d^2)) - (64\,C_F\,(-1+t)\,(6\,d^5 + d$$
$$\times (3374 - 368\,t) + d^3\,(652 - 58\,t) + d^4\,(-99+6\,t) + 8\,(-261+32\,t) + d^2\,(-2120+214\,t)))/(-84+73\,d - 21\,d^2 + 2\,d^3) + (32\,(C_A - 2$$
$$\times C_F)\,(-1+t)\,(-177\,d^6 + 6\,d^7 + d\,(160286 - 4410\,t) + d^5\,(2159 - 9\,t) - 40\,d^3\,(-1389+25\,t) + 80\,(-1062+31\,t) + d^4\,(-14275+156$$
$$\times t) + d^2\,(-127609+3033\,t)))/(-840+1318\,d - 805\,d^2 + 240\,d^3 - 35\,d^4 + 2\,d^5) - (16\,C_A\,(d\,(213284+254936\,t - 55420\,t^2) + d^3$$
$$\times (49343+45326\,t - 16909\,t^2) + d^5\,(861+267\,t - 408\,t^2) + 6\,d^6\,(-5+2\,t+3\,t^2) + 32\,(-4063 - 5206\,t + 917\,t^2) + d^4\,(-9214 - 6308$$
$$\times t + 3666\,t^2) + 8\,d^2\,(-17821 - 19232\,t + 5313\,t^2)))/((-4+d)^2\,(-42+47\,d - 17\,d^2 + 2\,d^3)) - (16\,C_F\,(d^6\,(243+2168\,t - 1691$$
$$\times t^2) + d^7\,(-9 - 60\,t + 69\,t^2) + 4\,d^5\,(-686 - 7277\,t + 4447\,t^2 + 12\,t^3) - 128\,(-630 - 9179\,t + 3115\,t^2 + 82\,t^3) - 4\,d^4\,(-4227 - 50286$$
$$\times t + 25915\,t^2 + 176\,t^3) + 4\,d^3\,(-15415 - 198612\,t + 89977\,t^2 + 1038\,t^3) + 16\,d\,(-9942 - 141011\,t + 52351\,t^2 + 1126\,t^3) - 8\,d^2$$
$$\times (-16683 - 227431\,t + 92720\,t^2 + 1532\,t^3)))/((-4+d)^2\,(-42+47\,d - 17\,d^2 + 2\,d^3)\,t) + ((-C_A + 2\,C_F)\,(9\,d^9\,(805 - 631\,t - 225$$
$$\times t^2 + 43\,t^3 + 8\,t^4) - 2\,d^8\,(135397 - 84999\,t - 36983\,t^2 + 6209\,t^3 + 1368\,t^4) - 2048\,(561744 - 75694\,t - 139390\,t^2 + 13653\,t^3 + 5262$$
$$\times t^4) + d^7\,(4339045 - 2234859\,t - 1166673\,t^2 + 176271\,t^3 + 44008\,t^4) - 4\,d^6\,(9866477 - 4230249\,t - 2621424\,t^2 + 362864$$
$$\times t^3 + 99864\,t^4) + 256\,d\,(11381020 - 1945608\,t - 2865944\,t^2 + 294961\,t^3 + 109030\,t^4) - 64\,d^2\,(50669002 - 10641039\,t - 12922019$$
$$\times t^2 + 1397323\,t^3 + 494026\,t^4) - 16\,d^4\,(52863551 - 15981438\,t - 13769839\,t^2 + 1659999\,t^3 + 528038\,t^4) + 4\,d^5\,(56449324 - 20302405$$
$$\times t - 14846598\,t^2 + 1908263\,t^3 + 568312\,t^4) + 16\,d^3\,(129983387 - 32915244\,t - 33519253\,t^2 + 3817778\,t^3 + 1284792\,t^4)))/((6 - 5$$
$$\times d + d^2)\,(28 - 15\,d + 2\,d^2)^2\,(24+8\,t - 4\,t^2 + d\,(-7 - 2\,t + t^2))))\,master[MI301pt1])/(128\,(-1+t)^2) + (C_F\,N^2\,(C_A\,t\,(9\,d^{11}\,(19+436$$
$$\times t - 773\,t^2 - 5192\,t^3 + 4753\,t^4 + 724\,t^5 + 33\,t^6) - 18\,d^{10}\,(313 + 11440\,t - 14735\,t^2 - 125324\,t^3 + 92943\,t^4 + 15292\,t^5 + 1639$$
$$\times t^6) - 30720\,(876 + 668866\,t + 275795\,t^2 - 5528709\,t^3 + 2827251\,t^4 - 204895\,t^5 + 180940\,t^6) + d^9\,(77057 + 4676180\,t - 4336927$$
$$\times t^2 - 47755888\,t^3 + 30060995\,t^4 + 5075804\,t^5 + 848667\,t^6) - 2\,d^8\,(273381 + 30690554\,t - 19737205\,t^2 - 295086080\,t^3 + 164110775$$
$$\times t^4 + 26972646\,t^5 + 6269177\,t^6) + 1536\,d\,(-24524 + 40620463\,t + 14760647\,t^2 - 328217041\,t^3 + 161598109\,t^4 - 6055060$$
$$\times t^5 + 10583730\,t^6) + 4\,d^7\,(473566 + 130270045\,t - 53517615\,t^2 - 1189240496\,t^3 + 605382490\,t^4 + 91401619\,t^5 + 28466167$$
$$\times t^6) - 256\,d^2\,(-697446 + 328619446\,t + 95753413\,t^2 - 2636473917\,t^3 + 1257344058\,t^4 - 2035927\,t^5 + 83078297\,t^6) -$$

$$8\,d^6\,(37494 + 377106649\,t - 81887231\,t^2 - 3295066056\,t^3 + 1583756642\,t^4 + 205221103\,t^5 + 85978951\,t^6)$$
$$+\,128\,d^3\,(-1553570 + 520256311\,t + 103028922\,t^2 - 4198709707\,t^3 + 1957981872\,t^4 + 67304114\,t^5 + 128221686\,t^6)$$
$$+\,16\,d^5\,(-1554383 + 761730823\,t - 41180284\,t^2 - 6423061412\,t^3 + 2993779937\,t^4 + 304541733\,t^5 + 178959986\,t^6)$$
$$-\,32\,d^4\,(-3238611 + 1075339321\,t + 89636978\,t^2 - 8827674818\,t^3 + 4079440051\,t^4 + 282927965\,t^5 + 258901634\,t^6))$$
$$+\,2\,C_F\,(-30720\,(-22050 + 434370\,t + 1540362\,t^2 - 2590639\,t^3 - 2781159\,t^4 + 5066957\,t^5 - 1432055\,t^6 + 10\,t^7)$$
$$+\,9\,d^{11}\,(-16 + 333\,t + 172\,t^2 - 2859\,t^3 - 584\,t^4 + 4527\,t^5 - 1684\,t^6 + 111\,t^7)$$
$$-\,6\,d^{10}\,(-1108 + 22853\,t + 22712\,t^2 - 190087\,t^3 - 86704\,t^4 + 288299\,t^5 - 116820\,t^6 + 5559\,t^7)$$
$$+\,512\,d\,(-3704400 + 74442654\,t + 244309327\,t^2 - 439242589\,t^3 - 478013543\,t^4 + 847517789\,t^5 - 256958670\,t^6 + 159660\,t^7)$$
$$+\,d^9\,(-138160 + 2837527\,t + 3945852\,t^2 - 22512713\,t^3 - 14470752\,t^4 + 33911941\,t^5 - 14318252\,t^6 + 496365\,t^7)$$
$$-\,256\,d^2\,(-9390135 + 191201668\,t + 583800846\,t^2 - 1131031291\,t^3 - 1237243458\,t^4 + 2138998840\,t^5 - 693970463\,t^6 + 1184293\,t^7)$$
$$-\,2\,d^8\,(-853780 + 17519331\,t + 30148470\,t^2 - 131575695\,t^3 - 103207780\,t^4 + 201675193\,t^5 - 85867998\,t^6 + 2165971\,t^7)$$
$$+\,128\,d^3\,(-14216421 + 291720203\,t + 830457359\,t^2 - 1752363535\,t^3 - 1906293278\,t^4 + 3226310223\,t^5 - 1116448498\,t^6 + 3734703\,t^7)$$
$$+\,4\,d^7\,(-3489064 + 71673744\,t + 143255651\,t^2 - 508691571\,t^3 - 453812920\,t^4 + 805078912\,t^5 - 337497027\,t^6 + 6129251\,t^7)$$
$$-\,8\,d^6\,(-9907732 + 203893170\,t + 456253483\,t^2 - 1371446853\,t^3 - 1333760260\,t^4 + 2257209742\,t^5 - 915715283\,t^6 + 11726117\,t^7)$$
$$-\,32\,d^4\,(-28553818 + 588002071\,t + 1558080619\,t^2 - 3630465070\,t^3 - 3875136680\,t^4 + 6465797345$$
$$\times\,t^5 - 2375914765\,t^6 + 13398418\,t^7) + 16\,d^5\,(-19962022 + 411304363\,t + 1007974965\,t^2 - 2638696560$$
$$\times\,t^3 - 2717543398\,t^4 + 4523245023\,t^5 - 1754064369\,t^6 + 15308062\,t^7)))\,master[MI302p])/(192\,(-5$$
$$+\,d)\,(-4 + d)^2\,(-3 + d)^2\,(-2 + d)\,(-7 + 2\,d)\,(-14 + 3\,d)\,(-10 + 3\,d)\,(-1 + t)^3\,t\,(1 + t)\,(6 - 10\,t + d\,(-1 + 3\,t)))$$
$$+\,(C_F\,N^2\,((-2\,I_2(R)\,(2\,d^4\,(4 + t + 3\,t^2) + 16\,(48 + 49\,t + 29\,t^2) - d^3\,(127 + 48\,t + 97\,t^2) - 2\,d\,(707 + 490\,t + 459\,t^2) + d^2\,(695 + 362\,t + 495$$
$$\times\,t^2)))/(168 - 262\,d + 111\,d^2 - 18\,d^3 + d^4) + ((C_A - 2\,C_F)\,(d^2\,(6681 + 2877\,t + 4823\,t^2 - 733\,t^3) + d^4\,(280 - 30\,t + 186\,t^2 - 44\,t^3) + 16$$
$$\times\,(483 + 413\,t + 373\,t^2 - 37\,t^3) + d^5\,(-16 + 7\,t - 10\,t^2 + 3\,t^3) + d^3\,(-1942 - 347\,t - 1350\,t^2 + 255\,t^3) + 2\,d\,(-5703 - 3735\,t - 4261$$
$$\times\,t^2 + 523\,t^3)))/((-4 + d)^2\,(6 - 5\,d + d^2)\,(1 + t)))\,master[MI303p])/(4\,(-1 + t)^2) - (C_F\,(C_A\,(1536 - 8576\,d + 24448\,d^2 - 21472$$
$$\times\,d^3 + 8114\,d^4 - 1363\,d^5 + 70\,d^6 + 3\,d^7) + 2\,(C_F\,(-159232 + 412704\,d - 416672\,d^2 + 217064\,d^3 - 63286\,d^4 + 10211\,d^5 - 810\,d^6 + 21$$
$$\times\,d^7) + 8\,(7168 - 13584\,d + 10548\,d^2 - 4312\,d^3 + 983\,d^4 - 119\,d^5 + 6\,d^6)\,I_2(R)\,(1 + 2\,n_f)))\,N^2\,master[MI401])/(8\,(-4 + d)^2$$
$$\times\,(-56 + 93\,d - 43\,d^2 + 6\,d^3)) + (C_F\,N^2\,(C_A\,(-9\,d^8\,t\,(16 - 21\,t + 5\,t^2) + d^6\,(4112 - 56808\,t + 54931\,t^2 - 11727\,t^3) + 12\,d^7\,(-16 + 374$$
$$\times\,t - 401\,t^2 + 91\,t^3) - 1536\,(-224 + 1092\,t - 1012\,t^2 + 157\,t^3) + 12\,d^5\,(-3110 + 32434\,t - 30049\,t^2 + 6033\,t^3) + 64\,d\,(-13996 + 77516$$
$$\times\,t - 72638\,t^2 + 12097\,t^3) + 16\,d^3\,(-34304 + 250432\,t - 231637\,t^2 + 42146\,t^3) - 16\,d^2\,(-59628 + 377462\,t - 352573\,t^2 + 61568\,t^3) - 4$$
$$\times\,d^4\,(-46504 + 398374\,t - 365896\,t^2 + 69625\,t^3)) - 2\,(16\,(-2 + d)^2\,(28 - 15\,d + 2\,d^2)\,I_2(R)\,(1 + 2\,n_f)\,(-1 + t^2)\,(d^2\,(34 - 31\,t) + 16$$
$$\times\,(9 - 8\,t) + 3\,d^3\,(-1 + t) + 2\,d\,(-62 + 55\,t)) + C_F\,(-1 + d)\,(d^6\,(548 - 2912\,t + 2269\,t^2 - 481\,t^3) + 3\,d^7\,(-8 + 32\,t - 25\,t^2 + t^3) - 512$$
$$\times\,(-336 + 2016\,t - 1892\,t^2 + 761\,t^3) + 4\,d^5\,(-1382 + 8722\,t - 7025\,t^2 + 2112\,t^3) + 64\,d\,(-4982 + 32370\,t - 30044\,t^2 + 12001\,t^3) - 4$$
$$\times\,d^4\,(-8016 + 55270\,t - 46500\,t^2 + 16223\,t^3) + 16\,d^3\,(-7205 + 50865\,t - 44618\,t^2 + 16780\,t^3) - 16\,d^2\,(-15946 + 109820\,t - 99631$$
$$\times\,t^2 + 39000\,t^3))))\,master[MI402p])/(16\,(-4 + d)^2\,(-2 + d)\,(-1 + d)\,(-7 + 2\,d)\,(-8 + 3\,d)\,(-1 + t)^2\,(1 + t)) - ((C_A - 2\,C_F)\,C_F\,N^2$$
$$\times\,(d^5\,(42538 + 115188\,t - 433388\,t^2 + 310772\,t^3 - 22118\,t^4) + d^7\,(282 + 896\,t - 3672\,t^2 + 2648\,t^3 - 154\,t^4) + d^8\,(-7 - 26\,t + 108$$
$$\times\,t^2 - 78\,t^3 + 3\,t^4) + 640\,(-990 - 4011\,t + 10219\,t^2 - 6951\,t^3 + 389\,t^4) + d^6\,(-4709 - 13510\,t + 53364\,t^2 - 38386\,t^3 + 2601\,t^4) - 32$$
$$\times\,d\,(-46047 - 155771\,t + 437639\,t^2 - 303951\,t^3 + 18394\,t^4) + 4\,d^4\,(-56872 - 151727\,t + 538895\,t^2 - 384877\,t^3 + 27285\,t^4) + 16$$
$$\times\,d^2\,(-88619 - 263469\,t + 809095\,t^2 - 569587\,t^3 + 36916\,t^4) - 8\,d^3\,(-92248 - 253619\,t + 842155\,t^2 - 598039\,t^3 + 40943\,t^4))$$
$$\times\,master[MI403p])/(32\,(-5 + d)\,(12 - 7\,d + d^2)\,(14 - 11\,d + 2\,d^2)\,(-1 + t)^2\,(6 - 10\,t + d\,(-1 + 3\,t))) + ((C_A - 2\,C_F)\,C_F\,N^2\,t^2\,(d^4$$
$$\times\,(22924 - 9934\,t - 6246\,t^2) + d^7\,(-7 + 6\,t + t^2) - 10\,d^6\,(-24 + 19\,t + 5\,t^2) - 64\,(-1650 - 1775\,t + 778\,t^2) + d^5\,(-3269 + 2054\,t + 799\,t^2)$$

$$
\begin{aligned}
&+8\,d^3\,(-11243+2331\,t+3404\,t^2)+16\,d\,(-14249-6746\,t+5646\,t^2)-8\,d^2\,(-24790-1911\,t+8478\,t^2))\,master[MI404p]) \\
&/(4\,(-4+d)^2\,(-3+d)\,(-2+d)\,(-7+2\,d)\,(-1+t)^2\,(6-10\,t+d\,(-1+3\,t)))-(C_F\,I_2(R)\,N^2\,(6\,d^5\,t-480\,(7+5\,t+t^2)-d^4\,(32+134 \\
&\times t+9\,t^2)+d^3\,(512+1108\,t+117\,t^2)-4\,d^2\,(688+1005\,t+128\,t^2)+d\,(5632+5992\,t+884\,t^2))\,master[MI406t])/((280-481 \\
&\times d+245\,d^2-47\,d^3+3\,d^4)\,t)+(C_F\,N^2\,((-4\,I_2(R)\,(6\,d^5\,(-1+t)^2-160\,(4+8\,t+6\,t^2+3\,t^3)-d^4\,(113-197\,t+107\,t^2+9\,t^3)+d^3 \\
&\times (791-1153\,t+757\,t^2+117\,t^3)-4\,d^2\,(615-676\,t+621\,t^2+128\,t^3)+4\,d\,(757-405\,t+835\,t^2+221\,t^3)))/(280-481\,d+245 \\
&\times d^2-47\,d^3+3\,d^4)+((-C_A+2\,C_F)\,(6\,d^7\,(-1+t)^2+d^3\,(84744-97060\,t+79828\,t^2-6376\,t^3)+d^5\,(2774-4300\,t+2650\,t^2-180 \\
&\times t^3)+1280\,(-120+71\,t-113\,t^2+8\,t^3)+d^6\,(-203+359\,t-197\,t^2+9\,t^3)-32\,d\,(-8622+6649\,t-8105\,t^2+611\,t^3)+d^4 \\
&\times (-20145+27053\,t-19071\,t^2+1475\,t^3)+4\,d^2\,(-51935+49617\,t-48819\,t^2+3841\,t^3)))/(-960+1592\,d-1030\,d^2+325\,d^3-50 \\
&\times d^4+3\,d^5))\,master[MI407p])/(4\,(-1+t)^2)+(2\,C_F\,I_2(R)\,N^2\,(4+t)\,(328+d^2\,(26-29\,t)-48\,t+d^3\,(-2+3\,t)+2\,d\,(-84+37\,t)) \\
&\times master[MI407t])/(280-481\,d+245\,d^2-47\,d^3+3\,d^4)+(C_F\,N^2\,((4\,I_2(R)\,(1+t)^2\,(d^2\,(-67+84\,t-29\,t^2)+d^3\,(5-8\,t+3\,t^2)-8 \\
&\times (55-53\,t+6\,t^2)+d\,(298-316\,t+74\,t^2)))/((280-481\,d+245\,d^2-47\,d^3+3\,d^4)\,(-1+t))+((-C_A+2\,C_F)\,(d^2\,(9402+1282 \\
&\times t+7318\,t^2-1042\,t^3)+d^4\,(354-142\,t+254\,t^2-50\,t^3)+d^5\,(-19+13\,t-13\,t^2+3\,t^3)-128\,(-93-58\,t-77\,t^2+8\,t^3)+8\,d \\
&\times (-2103-816\,t-1691\,t^2+206\,t^3)+d^3\,(-2597+331\,t-1947\,t^2+325\,t^3)))/(-960+1592\,d-1030\,d^2+325\,d^3-50\,d^4+3\,d^5)) \\
&\times master[MI408p])/(2\,(1+t))+((C_A-2\,C_F)\,C_F\,N^2\,(d^6\,(-9206+9279\,t-1205\,t^2+1061\,t^3-441\,t^4)+3\,d^7\,(98-105\,t+19\,t^2-19 \\
&\times t^3+7\,t^4)-512\,(8786-5994\,t-201\,t^2+100\,t^3+53\,t^4)+4\,d^5\,(29106-27480\,t+2463\,t^2-1968\,t^3+983\,t^4)+64\,d\,(135432-98504 \\
&\times t-1503\,t^2+798\,t^3+1213\,t^4)+16\,d^3\,(191690-158649\,t+4572\,t^2-3155\,t^3+3392\,t^4)-4\,d^4\,(196060-173268\,t+9799\,t^2-7226 \\
&\times t^3+4779\,t^4)-16\,d^2\,(437598-339421\,t+1770\,t^2-1277\,t^3+5574\,t^4))\,master[MI409p])/(8\,(448-632\,d+326\,d^2-73\,d^3+6\,d^4) \\
&\times (-1+t)^2\,(24+8\,t-4\,t^2+d\,(-7-2\,t+t^2)))+(C_F\,(-C_A+2\,C_F)\,N^2\,(d^6\,(-1066+3411\,t-3441\,t^2+1191\,t^3-95\,t^4)+3\,d^7\,(14-45 \\
&\times t+45\,t^2-15\,t^3+t^4)-512\,(1196-2482\,t+1804\,t^2-563\,t^3+66\,t^4)+2\,d^5\,(6098-18721\,t+18620\,t^2-6554\,t^3+581\,t^4)+64\,d \\
&\times (16912-37640\,t+29614\,t^2-9675\,t^3+1082\,t^4)-4\,d^4\,(20186-57775\,t+55425\,t^2-19516\,t^3+1862\,t^4)+8\,d^3\,(41018-107859 \\
&\times t+97985\,t^2-34043\,t^3+3444\,t^4)-16\,d^2\,(50228-121124\,t+102817\,t^2-34826\,t^3+3710\,t^4))\,master[MI410p])/(4\,(-4+d)^2 \\
&\times (-112+130\,d-49\,d^2+6\,d^3)\,(-1+t)\,(24+8\,t-4\,t^2+d\,(-7-2\,t+t^2)))+((C_A-2\,C_F)\,C_F\,N^2\,(3\,d^7\,(42-51\,t+8\,t^2+t^3)-512 \\
&\times (2538-2210\,t+169\,t^2+66\,t^3)-d^6\,(3446-4019\,t+542\,t^2+95\,t^3)+2\,d^5\,(19776-22084\,t+2639\,t^2+581\,t^3)-8\,d^4\,(31104-33150 \\
&\times t+3570\,t^2+931\,t^3)+64\,d\,(39320-36048\,t+2981\,t^2+1082\,t^3)+8\,d^3\,(116196-117910\,t+11557\,t^2+3444\,t^3)-16\,d^2 \\
&\times (128942-124409\,t+11172\,t^2+3710\,t^3))\,master[MI411p])/(8\,(-1344+2344\,d-1610\,d^2+545\,d^3-91\,d^4+6\,d^5)\,(24+8\,t-4 \\
&\times t^2+d\,(-7-2\,t+t^2)))+(C_F\,N^2\,(C_A\,t\,(d^2\,(-4664-87668\,t+43792\,t^2+4524\,t^3-816\,t^4)+d^4\,(-164-2651\,t+1377\,t^2+311 \\
&\times t^3-25\,t^4)+d^5\,(8+107\,t-93\,t^2-23\,t^3+t^4)-96\,(63+1105\,t-991\,t^2-34\,t^3+11\,t^4)+2\,d^3\,(630+11289\,t-5039\,t^2-841\,t^3+105 \\
&\times t^4)+8\,d\,(1056+19773\,t-12764\,t^2-753\,t^3+188\,t^4))-2\,C_F\,(d^2\,(-2920+46664\,t+64988\,t^2-58416\,t^3+9348\,t^4-816\,t^5)+d^4 \\
&\times (-86+1400\,t+1501\,t^2-1915\,t^3+277\,t^4-25\,t^5)+d^5\,(4-64\,t-37\,t^2+107\,t^3-11\,t^4+t^5)-32\,(147-2226\,t-3139\,t^2+2809 \\
&\times t^3-396\,t^4+33\,t^5)+2\,d^3\,(358-5810\,t-7543\,t^2+7353\,t^3-1183\,t^4+105\,t^5)+8\,d\,(735-11469\,t-16268\,t^2+14501\,t^3-2195 \\
&\times t^4+188\,t^5)))\,master[MI501p])/(8\,(-7+2\,d)\,(6-5\,d+d^2)\,(-1+t)^2\,t\,(1+t))-(C_F\,(-9+2\,d)\,N^2\,(C_A\,t\,(-3\,d^5\,(-15+14\,t+t^2)+d^4 \\
&\times (-1085+690\,t+11\,t^2)-96\,(488-469\,t+37\,t^2)+24\,d\,(2825-2093\,t+172\,t^2)+d^3\,(9286-5264\,t+202\,t^2)-4\,d^2\,(9171-5611 \\
&\times t+400\,t^2))+2\,C_F\,(d^5\,(16+19\,t-38\,t^2+3\,t^3)-d^4\,(344+675\,t-710\,t^2+75\,t^3)-32\,(588+1406\,t-1085\,t^2+99\,t^3)+8\,d\,(2940+7265 \\
&\times t-5585\,t^2+564\,t^3)-4\,d^2\,(2920+7217\,t-5589\,t^2+612\,t^3)+d^3\,(2864+6682\,t-5568\,t^2+630\,t^3)))\,master[MI502p])/(8\,(-4+d)^2 \\
&\times (-7+2\,d)\,(6-5\,d+d^2)\,t\,(1+t))+((C_A-2\,C_F)\,C_F\,(-176+104\,d-20\,d^2+d^3)\,N^2\,t^2\,master[MI601])/(16\,(-7+2\,d))+((C_A-2 \\
&\times C_F)\,C_F\,N^2\,(-20\,d^2\,(-1+t)^2+d^3\,(-1+t)^2-16\,(10-21\,t+11\,t^2)+4\,d\,(25-51\,t+26\,t^2))\,master[MI601p])/(16\,(-7+2\,d)))
\end{aligned}
$$

(C.3)

Calculation of form factors

D.1. Tensor coefficients for form factors

When we use the projection method to calculate the axial form factors as well as vector form factors we need the tensor coefficients as described in subsec. 2.1.3. In this appendix we state the tensor coefficients applied to this calculation:

$$
B_1 = \frac{1}{(2\,(-2+d)\,t^2\,(-4\,m^2+t)^2)} \bigg((-2+d)\,t^2\,S_3 - 4\,(-1+d)\,m^4\,(S_1 - S_2 - S_3 + S_4)
$$
$$
+ 2\,m^2\,t\left((-1+d)\,S_1 - 2\,(-2+d)\,S_3 + (-1+d)\,S_4\right) - 4\,m^3\,t\,S_6 + m\,t^2\,S_6 \bigg),
$$

$$
B_2 = \frac{1}{(2\,(-2+d)\,t^2\,(-4\,m^2+t)^2)} \bigg((-2+d)\,t^2\,S_4 + 4\,(-1+d)\,m^4\,(S_1 - S_2 - S_3 + S_4)
$$
$$
+ 2\,m^2\,t\left((-1+d)\,S_2 + (-1+d)\,S_3 - 2\,(-2+d)\,S_4\right) - 4\,m^3\,t\,S_5 + m\,t^2\,S_5 \bigg),
$$

$$
B_3 = \frac{1}{(2\,(-2+d)\,t^2\,(-4\,m^2+t)^2)} \bigg((-2+d)\,t^2\,S_1 - 2\,m^2\,t\left(2\,(-2+d)\,S_1 - (-1+d)\,(S_2 + S_3)\right)
$$
$$
+ 4\,(-1+d)\,m^4\,(S_1 - S_2 - S_3 + S_4) - 4\,m^3\,t\,S_5 + m\,t^2\,S_5 \bigg),
$$

$$
B_4 = \frac{1}{(2\,(-2+d)\,t^2\,(-4\,m^2+t)^2)} \bigg((-2+d)\,t^2\,S_2 - 4\,(-1+d)\,m^4\,(S_1 - S_2 - S_3 + S_4)
$$
$$
+ 2\,m^2\,t\left((-1+d)\,S_1 - 2\,(-2+d)\,S_2 + (-1+d)\,S_4\right) - 4\,m^3\,t\,S_6 + m\,t^2\,S_6 \bigg),
$$

$$
B_5 = \frac{2\,m\,(S_2 + S_3) + t\,S_5 - 2\,m^2\,(S_5 + S_6)}{4\,(-2+d)\,t\,(-4\,m^2+t)}, \qquad B_6 = \frac{2\,m\,(S_1 + S_4) + t\,S_6 - 2\,m^2\,(S_5 + S_6)}{4\,(-2+d)\,t\,(-4\,m^2+t)}.
$$

Here t is the Mandelstam variable

$$(p_2 - q_1)^2 = t, \qquad q_1^2 = p_2^2 = m^2, \tag{D.1}$$

and m is the mass of the massive quark and d the space-time dimension.

D.2. Export File for Reduze2

In order to choose our own master integrals in Reduze2[25] we have to write the following export file:

```
jobs:

  - select_reductions:
      input_file: "integrals_to_be_reduced"
      output_file: "reduced_integrals.out"

  - select_reductions:
      input_file:  "preferred_integrals.in"
      output_file: "preferred_integrals.out"

  - reduce_files:
      equation_files:
        - "reduced_integrals.out"
        - "preferred_integrals.out"
      output_file: "transf.out"
      preferred_masters_file: "preferred_integrals.in"

  - export:
      conditional: false
      input_file:  "transf.out"
      output_format: mma
      preferred_masters_file: "preferred_integrals.in"
```

We write our own master integrals in the file *"preferred_integrals.in"*. In the part *reduce_files*, the equations are solved by selected integrals.

Tensor coefficients for double boxes

Using the projection method for the calculation of double boxes we need the tensor coefficients as described in subsec. 2.1.3. In this appendix we state the tensor coefficients applied to this calculation:

$$
\begin{aligned}
B_1 = -\Bigg(& 8\,m_t^3\,\Big((d-2)\,s\,((432+76\,d+4\,d^2+5\,d^3)\,S_3 - (24+6\,d+d^2)\,S_7) + (-7+d)\,t\,((192+56\,d+14\,d^2+5\,d^3)\,S_3 \\
& - (24+6\,d+d^2)\,S_7)\Big) + (-7+d)\,t\,(s+t)\,((4416-1408\,d+940\,d^2-100\,d^3+15\,d^4)\,S_1 - 10\,d^3\,(4\,S_4+S_5) \\
& - 2\,d\,(224\,S_4+80\,S_5-8\,S_8-3\,S_9)+24\,(-24\,S_5+S_9)+d^2\,(144\,S_4-36\,S_5+8\,S_8+S_9)) - m_t^2\,\Big(t\,((-30912+14272\,d \\
& - 7988\,d^2+1640\,d^3-205\,d^4+15\,d^5)\,S_1 - 40\,(96+96\,d+22\,d^2+3\,d^3)\,S_4 + 4032\,S_5 + 544\,d\,S_5 + 92\,d^2\,S_5 + 34\,d^3\,S_5 \\
& - 10\,d^4\,S_5 + 960\,S_8 + 240\,d\,S_8 + 40\,d^2\,S_8 - 168\,S_9 - 18\,d\,S_9 - d^2\,S_9 + d^3\,S_9) + (d-2)\,s\,((11136-3728\,d+1500\,d^2 \\
& - 160\,d^3+15\,d^4)\,S_1 - 2\,(528+100\,d+8\,d^2+5\,d^3)\,S_5 + (24+6\,d+d^2)\,S_9)\Big) + m_t\,\Big(t\,((-26112+11840\,d-8896\,d^2 \\
& + 1612\,d^3-260\,d^4+15\,d^5)\,S_2 + 1920\,S_6 + 1312\,d\,S_6 + 328\,d^2\,S_6 + 88\,d^3\,S_6 - 10\,d^4\,S_6 - 8\,(-7+d)\,t\,((192+56\,d+14\,d^2 \\
& + 5\,d^3)\,S_3 - (24+6\,d+d^2)\,S_7) - 288\,S_{10} - 48\,d\,S_{10} - 6\,d^2\,S_{10} + d^3\,S_{10}) + s\,((-14592+16160\,d-7336\,d^2+1852\,d^3-230\,d^4 \\
& + 15\,d^5)\,S_2 - (d-2)\,(2\,(240+28\,d-4\,d^2+5\,d^3)\,S_6 - (24+6\,d+d^2)\,S_{10}) + t\,((14208+1264\,d+540\,d^2+208\,d^3-25\,d^4)\,S_3 \\
& - (24+6\,d+d^2)\,(2\,(18+d)\,S_7 - S_{11})))\Big)\Bigg) \Big/ \Bigg(32\,(-2520+2754\,d-1175\,d^2+245\,d^3-25\,d^4+d^5)\,s\,t^2\,(s+t)\,(-m_t^2+s+t)\Bigg)
\end{aligned}
$$

$$
\begin{aligned}
B_2 = \Bigg(& -8\,m_t^4\,s\,\Big((d-2)\,s\,((32+100\,d-28\,d^2+5\,d^3)\,S_3 - (8-2\,d+d^2)\,S_7) + t\,((-704-728\,d+56\,d^2-68\,d^3+5\,d^4)\,S_3 \\
& + (176+8\,d+14\,d^2-d^3)\,S_7)\Big) - m_t^3\,\Big(-8\,(d-2)\,t^2\,((432+76\,d+4\,d^2+5\,d^3)\,S_4 - (24+6\,d+d^2)\,S_8) + s\,t\,((26112-11840\,d \\
& + 8896\,d^2-1612\,d^3+260\,d^4-15\,d^5)\,S_1 - 8\,(-1504-280\,d-32\,d^2-36\,d^3+5\,d^4)\,S_4 - 2816\,S_5 - 1520\,d\,S_5 - 128\,d^3\,S_5 \\
& + 10\,d^4\,S_5 - 1664\,S_8 - 64\,d\,S_8 - 48\,d^2\,S_8 + 8\,d^3\,S_8 + 176\,S_9 + 8\,d\,S_9 + 14\,d^2\,S_9 - d^3\,S_9) - (d-2)\,s^2\,((7296-4432\,d \\
& + 1452\,d^2-200\,d^3+15\,d^4)\,S_1 - 2\,(64+92\,d-24\,d^2+5\,d^3)\,S_5 + (8-2\,d+d^2)\,S_9)\Big) - m_t\,(s+t)\,\Big(8\,(d-2)\,t^2\,((432+76\,d+4\,d^2 \\
& + 5\,d^3)\,S_4 - (24+6\,d+d^2)\,S_8) + s\,t\,((-26112+11840\,d-8896\,d^2+1612\,d^3-260\,d^4+15\,d^5)\,S_1 + 8\,(-2624+524\,d-236\,d^2 \\
& + 15\,d^3)\,S_4 + 2816\,S_5 + 1520\,d\,S_5 + 128\,d^3\,S_5 - 10\,d^4\,S_5 + 1888\,S_8 + 32\,d\,S_8 - 8\,d^2\,S_8 - 176\,S_9 - 8\,d\,S_9 - 14\,d^2\,S_9 + d^3\,S_9) \\
& + (d-2)\,s^2\,((7296-4432\,d+1452\,d^2-200\,d^3+15\,d^4)\,S_1 - 2\,(64+92\,d-24\,d^2+5\,d^3)\,S_5 + (8-2\,d+d^2)\,S_9)\Big) \\
& + m_t^2\,\Big(8\,(d-2)\,s^3\,((32+100\,d-28\,d^2+5\,d^3)\,S_3 - (8-2\,d+d^2)\,S_7) - (d-2)\,t^2\,((11136-3728\,d+1500\,d^2-160\,d^3+15\,d^4)\,S_2
\end{aligned}
$$

$$-2\left(528+100\,d+8\,d^2+5\,d^3\right)S_6+\left(24+6\,d+d^2\right)S_{10}+s^2\left(\left(13824-17312\,d+8520\,d^2-2140\,d^3+270\,d^4-15\,d^5\right)S_2\right.$$

$$+(d-2)\left(2\left(-32+116\,d-36\,d^2+5\,d^3\right)S_6-\left(8-2\,d+d^2\right)S_{10}\right)+t\left(\left(-7296-9616\,d+2788\,d^2-1192\,d^3+105\,d^4\right)S_3\right.$$

$$+2\left(688+100\,d+72\,d^2-7\,d^3\right)S_7-\left(8-2\,d+d^2\right)S_{11}\right)+s\,t\left(\left(51456-32064\,d+17408\,d^2-3720\,d^3+520\,d^4-30\,d^5\right)S_2\right.$$

$$+2\left(2\left(-816-488\,d+56\,d^2-54\,d^3+5\,d^4\right)S_6+\left(192+8\,d+10\,d^2-d^3\right)S_{10}\right)+t\left(\left(-2176-9296\,d+2108\,d^2-928\,d^3+95\,d^4\right)S_3\right.$$

$$+\left(1888+56\,d+112\,d^2-26\,d^3\right)S_7+\left(24+6\,d+d^2\right)S_{11}\right)\right)+(s+t)\left(\left(d-2\right)t^2\left(\left(11136-3728\,d+1500\,d^2-160\,d^3+15\,d^4\right)S_2\right.\right.$$

$$-2\left(528+100\,d+8\,d^2+5\,d^3\right)S_6+\left(24+6\,d+d^2\right)S_{10}+s^2\left(\left(-13824+17312\,d-8520\,d^2+2140\,d^3-270\,d^4+15\,d^5\right)S_2\right.$$

$$-(d-2)\left(2\left(-32+116\,d-36\,d^2+5\,d^3\right)S_6-\left(8-2\,d+d^2\right)S_{10}\right)+t\left(\left(7424-3600\,d+1212\,d^2-160\,d^3+15\,d^4\right)S_3\right.$$

$$-2\left(32+100\,d-28\,d^2+5\,d^3\right)S_7+\left(8-2\,d+d^2\right)S_{11}\right)-s\,t\left(\left(67584-44672\,d+21120\,d^2-4600\,d^3+600\,d^4-30\,d^5\right)S_2\right.$$

$$+2\left(2\left(-1152-440\,d+140\,d^2-66\,d^3+5\,d^4\right)S_6+\left(192+8\,d+10\,d^2-d^3\right)S_{10}\right)+t\left(\left(3456-3472\,d+1660\,d^2-384\,d^3+55\,d^4\right)S_3\right.$$

$$\left.\left.-2\left(-240+4\,d+9\,d^3\right)S_7+\left(24+6\,d+d^2\right)S_{11}\right)\right)\right)\bigg/\left(32\left(-2520+2754\,d-1175\,d^2+245\,d^3-25\,d^4+d^5\right)s^2\,t^2\,(s+t)\left(-m_t^2+s+t\right)^2\right)$$

$$B_3=\left(-320\,m_t^6\left(\left(-7+d\right)t\left(4\left(13+3\,d^2\right)S_3-\left(4+3\,d\right)S_7\right)+(d-2)\,s\left(\left(142-15\,d+12\,d^2\right)S_3-\left(4+3\,d\right)S_7\right)\right)\right.$$

$$+5\,m_t\,t\,(s+t)\left(8\,t\left(3\left(-1344-200\,d-42\,d^2-21\,d^3+5\,d^4\right)S_1+\left(576+232\,d-90\,d^2+45\,d^3\right)S_4+952\,S_5+32\,d\,S_5+144\,d^2\,S_5\right.\right.$$

$$-24\,d^3\,S_5-60\,S_8-4\,d\,S_8-9\,d^2\,S_8-28\,S_9-17\,d\,S_9+3\,d^2\,S_9\right)+s\left(3\left(-14208-1264\,d-540\,d^2-208\,d^3+25\,d^4\right)S_1\right.$$

$$-4\left(2240-2028\,d+748\,d^2-177\,d^3+15\,d^4\right)S_4+9920\,S_5+320\,d\,S_5+1092\,d^2\,S_5-102\,d^3\,S_5+560\,S_8-192\,d\,S_8-68\,d^2\,S_8$$

$$\left.+12\,d^3\,S_8-368\,S_9-86\,d\,S_9+9\,d^2\,S_9\right)\right)+40\,m_t^5\left(t\left(3\left(-1344-200\,d-42\,d^2-21\,d^3+5\,d^4\right)S_1-40\left(4+3\,d\right)^2\,S_4+952\,S_5\right.\right.$$

$$+32\,d\,S_5+144\,d^2\,S_5-24\,d^3\,S_5+160\,S_8+120\,d\,S_8-28\,S_9-17\,d\,S_9+3\,d^2\,S_9\right)+(d-2)\,s\left(3\left(432+76\,d+4\,d^2+5\,d^3\right)S_1\right.$$

$$\left.+\left(-316+6\,d-24\,d^2\right)S_5+\left(4+3\,d\right)S_9\right)\right)-5\,m_t^3\left(8\,t^2\left(6\left(-1344-200\,d-42\,d^2-21\,d^3+5\,d^4\right)S_1+\left(-64-728\,d-450\,d^2+45\,d^3\right)S_4\right.\right.$$

$$+1904\,S_5+64\,d\,S_5+288\,d^2\,S_5-48\,d^3\,S_5+100\,S_8+116\,d\,S_8-9\,d^2\,S_8-56\,S_9-34\,d\,S_9+6\,d^2\,S_9\right)+s\,t\left(3\left(-31872-624\,d-332\,d^2\right.\right.$$

$$-424\,d^3+105\,d^4\right)S_1-320\left(4+3\,d\right)^2\,S_4+22592\,S_5-2048\,d\,S_5+2676\,d^2\,S_5-486\,d^3\,S_5+1952\,S_8+528\,d\,S_8+48\,d^2\,S_8-656\,S_9$$

$$\left.-238\,d\,S_9+57\,d^2\,S_9\right)+8\,(d-2)\,s^2\left(3\left(432+76\,d+4\,d^2+5\,d^3\right)S_1+\left(-316+6\,d-24\,d^2\right)S_5+\left(4+3\,d\right)S_9\right)\right)$$

$$+40\,m_t^4\left(8\,(d-2)\,s^2\left(\left(142-15\,d+12\,d^2\right)S_3-\left(4+3\,d\right)S_7\right)+t\left(-3\left(-704-728\,d+56\,d^2-68\,d^3+5\,d^4\right)S_2-600\,S_6+16\,d\,S_6\right.\right.$$

$$-306\,d^2\,S_6+24\,d^3\,S_6+16\left(-7+d\right)t\left(4\left(13+3\,d^2\right)S_3-\left(4+3\,d\right)S_7\right)+48\,S_{10}+32\,d\,S_{10}-3\,d^2\,S_{10}\right)+s\left(-3\left(-64-168\,d\right.\right.$$

$$+156\,d^2-38\,d^3+5\,d^4\right)S_2+(d-2)\left(\left(220-78\,d+24\,d^2\right)S_6-\left(4+3\,d\right)S_{10}\right)+t\left(2\left(-4624+872\,d-738\,d^2+99\,d^3\right)S_3\right.$$

$$\left.\left.+\left(576+218\,d-27\,d^2\right)S_7-\left(4+3\,d\right)S_{11}\right)\right)\right)-m_t^2\left(5\,t^2\left(-3\left(-2176-9296\,d+2108\,d^2-928\,d^3+95\,d^4\right)S_2-7648\,S_6+3448\,d\,S_6\right.\right.$$

$$-3792\,d^2\,S_6+414\,d^3\,S_6+64\left(-7+d\right)t\left(4\left(13+3\,d^2\right)S_3-\left(4+3\,d\right)S_7\right)+184\,S_{10}+398\,d\,S_{10}-51\,d^2\,S_{10}\right)-20\,s^2\left(6\left(-64-168\,d\right.\right.$$

$$+156\,d^2-38\,d^3+5\,d^4\right)S_2-2\,(d-2)\left(\left(220-78\,d+24\,d^2\right)S_6-\left(4+3\,d\right)S_{10}\right)+t\left(\left(5552+2844\,d-112\,d^2+201\,d^3-15\,d^4\right)S_3\right.$$

$$\left.+\left(-604-72\,d-71\,d^2+3\,d^3\right)S_7+2\left(4+3\,d\right)S_{11}\right)\right)+s\,t\left(-15\left(-7296-9616\,d+2788\,d^2-1192\,d^3+105\,d^4\right)S_2\right.$$

$$+5\left(2\left(-6160+2564\,d-2160\,d^2+213\,d^3\right)S_6+\left(536+314\,d-45\,d^2\right)S_{10}\right)+t\left(\left(-206784-54640\,d-17820\,d^2-1680\,d^3+525\,d^4\right)S_3\right.$$

$$\left.\left.+10\left(2392+444\,d+126\,d^2-21\,d^3\right)S_7+\left(-16-170\,d+15\,d^2\right)S_{11}\right)\right)+t\,(s+t)\left(-5\,t\left(3\left(3456-3472\,d+1660\,d^2-384\,d^3+55\,d^4\right)S_2\right.\right.$$

$$+\left(2848-3320\,d+1344\,d^2-222\,d^3\right)S_6+\left(200-142\,d+27\,d^2\right)S_{10}\right)+s\left(15\left(7424-3600\,d+1212\,d^2-160\,d^3+15\,d^4\right)S_2\right.$$

$$-5\left(2\left(1664-276\,d+84\,d^2+15\,d^3\right)S_6-\left(88+42\,d+3\,d^2\right)S_{10}\right)+t\left(\left(72256-69360\,d+28740\,d^2-5760\,d^3+525\,d^4\right)S_3\right.$$

$$\left.\left.\left.\left.-10\left(-344+364\,d-138\,d^2+21\,d^3\right)S_7+\left(144-50\,d+15\,d^2\right)S_{11}\right)\right)\right)\right)$$

$$\Big/\left(480\left(-2520+2754\,d-1175\,d^2+245\,d^3-25\,d^4+d^5\right)s\,t^2\,(s+t)\left(-m_t^2+s+t\right)^2\right)$$

$$B_4=\left(-80\left(4+3\,d\right)m_t^5\,s\left(\left(4+3\,d\right)S_3-S_7\right)+2\,m_t^4\left(-8\,(d-2)\,t\left(\left(142-15\,d+12\,d^2\right)S_4-\left(4+3\,d\right)S_8\right)+s\left(15\left(96+96\,d+22\,d^2+3\,d^3\right)S_1\right.\right.\right.$$

$$\left.\left.-32\left(-91+13\,d-21\,d^2+3\,d^3\right)S_4-\left(4+3\,d\right)\left(10\left(8+3\,d\right)S_5-8\left(-7+d\right)S_8-5\,S_9\right)\right)\right)+\left(-7+d\right)s\,(s+t)^2\left(15\,d^3\left(2\,S_1-S_4\right)+368\,S_4\right.$$

$$\left.-3\,d^2\left(36\,S_1-36\,S_4+11\,S_5-S_8\right)-2\left(92\,S_5-2\,S_8+S_9\right)+d\left(336\,S_1-340\,S_4+90\,S_5-8\,S_8+3\,S_9\right)\right)$$

$$
\begin{aligned}
&- m_t^2 (s+t) \Big(-16 (d-2) t ((142 - 15 d + 12 d^2) S_4 - (4 + 3 d) S_8) + s (6 (480 + 88 d + 292 d^2 - 38 d^3 + 5 d^4) S_1 \\
&\quad - 8 (-896 + 86 d - 99 d^2 + 15 d^3) S_4 + 648 S_5 - 1534 d S_5 + 141 d^2 S_5 - 33 d^3 S_5 - 616 S_8 - 164 d S_8 + 36 d^2 S_8 + 54 S_9 + 7 d S_9 \\
&\quad + 3 d^2 S_9)) + 2 m_t^3 \Big(2 s^2 (20 (4 + 3 d)^2 S_3 + (-38 - 87 d + 3 d^2) S_7) + (d-2) t (3 (432 + 76 d + 4 d^2 + 5 d^3) S_2 + (-316 + 6 d - 24 d^2) S_6 \\
&\quad + (4 + 3 d) S_{10}) + s (3 (-1504 - 280 d - 32 d^2 - 36 d^3 + 5 d^4) S_2 + 792 S_6 + 32 d S_6 + 234 d^2 S_6 - 24 d^3 S_6 - 48 S_{10} - 32 d S_{10} + 3 d^2 S_{10} \\
&\quad + t ((64 + 728 d + 450 d^2 - 45 d^3) S_3 + 2 (-78 - 97 d + 18 d^2) S_7 - (4 + 3 d) S_{11}))\Big) - m_t (s+t) \Big((14 - 9 d + d^2) s^2 ((160 - 42 d + 15 d^2) S_3 \\
&\quad + (2 - 3 d) S_7) + 2 (d-2) t (3 (432 + 76 d + 4 d^2 + 5 d^3) S_2 + (-316 + 6 d - 24 d^2) S_6 + (4 + 3 d) S_{10}) + 2 s (3 (-2624 + 524 d - 236 d^2 \\
&\quad + 15 d^3) S_2 + 1212 S_6 - 112 d S_6 + 183 d^2 S_6 - 15 d^3 S_6 - 48 S_{10} - 32 d S_{10} + 3 d^2 S_{10} - t ((576 + 232 d - 90 d^2 + 45 d^3) S_3 \\
&\quad + (-4 + 74 d - 36 d^2) S_7 + (4 + 3 d) S_{11}))\Big)\Big) \Big/ \Big(24 (-2520 + 2754 d - 1175 d^2 + 245 d^3 - 25 d^4 + d^5) s^2 t (s+t) (-m_t^2 + s + t)^2 \Big)
\end{aligned}
$$

$$
\begin{aligned}
B_5 = {}& \Big(8 m_t^3 ((-7 + d) t (4 (17 + 3 d + 3 d^2) S_3 - (8 + 3 d) S_7) + (d-2) s ((158 - 3 d + 12 d^2) S_3 - (8 + 3 d) S_7)) + (-7 + d) t (s+t) \Big(3 (288 \\
&+ 80 d + 18 d^2 + 5 d^3) S_1 - 2 (184 - 90 d + 33 d^2) S_4 - 200 S_5 - 48 d S_5 - 24 d^2 S_5 - 4 S_8 + 18 d S_8 + 8 S_9 + 3 d S_9\Big) \\
&+ m_t^2 \Big(t ((6048 + 816 d + 138 d^2 + 51 d^3 - 15 d^4) S_1 + 40 (32 + 36 d + 9 d^2) S_4 - 1400 S_5 - 136 d S_5 - 120 d^2 S_5 + 24 d^3 S_5 - 320 S_8 \\
&- 120 d S_8 + 56 S_9 + 13 d S_9 - 3 d^2 S_9) - (d-2) s (3 (528 + 100 d + 8 d^2 + 5 d^3) S_1 - 2 (190 + 9 d + 12 d^2) S_5 + (8 + 3 d) S_9)\Big) \\
&+ m_t \Big(t (3 (-1408 - 760 d - 64 d^3 + 5 d^4) S_2 + 696 S_6 + 304 d S_6 + 282 d^2 S_6 - 24 d^3 S_6 - 8 (-7 + d) t (4 (17 + 3 d + 3 d^2) S_3 \\
&- (8 + 3 d) S_7) - 96 S_{10} - 28 d S_{10} + 3 d^2 S_{10}) + s (3 (-128 - 120 d + 140 d^2 - 34 d^3 + 5 d^4) S_2 \\
&- (d-2) (2 (94 - 27 d + 12 d^2) S_6 - (8 + 3 d) S_{10}) + t ((4960 + 160 d + 546 d^2 - 51 d^3) S_3 - (8 + 3 d) (2 (18 + d) S_7 - S_{11})))\Big)\Big) \\
& \Big/ \Big(48 (-2520 + 2754 d - 1175 d^2 + 245 d^3 - 25 d^4 + d^5) s t^2 (s+t) (-m_t^2 + s + t) \Big)
\end{aligned}
$$

$$
\begin{aligned}
B_6 = {}& \Big(8 m_t^4 s \Big((d-2) s ((110 - 39 d + 12 d^2) S_3 + (4 - 3 d) S_7) + t ((-300 + 8 d - 153 d^2 + 12 d^3) S_3 + (12 + 40 d - 3 d^2) S_7)\Big) \\
&+ m_t^3 \Big(-8 (d-2) t^2 ((158 - 3 d + 12 d^2) S_4 - (8 + 3 d) S_8) + s t (3 (960 + 656 d + 164 d^2 + 44 d^3 - 5 d^4) S_1 + (3168 + 128 d \\
&+ 936 d^2 - 96 d^3) S_4 - 696 S_5 - 304 d S_5 - 282 d^2 S_5 + 24 d^3 S_5 - 288 S_8 - 224 d S_8 + 24 d^2 S_8 + 12 S_9 + 40 d S_9 - 3 d^2 S_9) \\
&- (d-2) s^2 (3 (240 + 28 d - 4 d^2 + 5 d^3) S_1 - 2 (94 - 27 d + 12 d^2) S_5 + (-4 + 3 d) S_9)\Big) + m_t (s+t) \Big(8 (d-2) t^2 ((158 \\
&- 3 d + 12 d^2) S_4 - (8 + 3 d) S_8) + s t (3 (-960 - 656 d - 164 d^2 - 44 d^3 + 5 d^4) S_1 + (-4848 + 448 d - 732 d^2 + 60 d^3) S_4 \\
&+ 696 S_5 + 304 d S_5 + 282 d^2 S_5 - 24 d^3 S_5 + 456 S_8 + 116 d S_8 - 12 d^2 S_8 - 12 S_9 - 40 d S_9 + 3 d^2 S_9) + (d-2) s^2 (3 (240 + 28 d \\
&- 4 d^2 + 5 d^3) S_1 - 2 (94 - 27 d + 12 d^2) S_5 + (-4 + 3 d) S_9)\Big) + m_t^2 \Big(-8 (d-2) s^3 ((110 - 39 d + 12 d^2) S_3 + (4 - 3 d) S_7) \\
&+ (d-2) t^2 (3 (528 + 100 d + 8 d^2 + 5 d^3) S_2 - 2 (190 + 9 d + 12 d^2) S_6 + (8 + 3 d) S_{10}) + s^2 (3 (64 - 264 d + 188 d^2 - 46 d^3 + 5 d^4) S_2 \\
&- (d-2) (2 (142 - 63 d + 12 d^2) S_6 + (4 - 3 d) S_{10}) + t ((6160 - 2564 d + 2160 d^2 - 213 d^3) S_3 + (-84 - 430 d + 36 d^2) S_7 \\
&+ (-4 + 3 d) S_{11})) - s t (-6 (-816 - 488 d + 56 d^2 - 54 d^3 + 5 d^4) S_2 - 1488 S_6 + 400 d S_6 - 564 d^2 S_6 + 48 d^3 S_6 + 48 S_{10} + 68 d S_{10} \\
&- 6 d^2 S_{10} + t ((-3824 + 1724 d - 1896 d^2 + 207 d^3) S_3 + (228 + 430 d - 72 d^2) S_7 + (8 + 3 d) S_{11}))\Big) - (s+t) \Big((d-2) t^2 (3 (528 \\
&+ 100 d + 8 d^2 + 5 d^3) S_2 - 2 (190 + 9 d + 12 d^2) S_6 + (8 + 3 d) S_{10}) + s^2 (3 (64 - 264 d + 188 d^2 - 46 d^3 + 5 d^4) S_2 \\
&- (d-2) (2 (142 - 63 d + 12 d^2) S_6 + (4 - 3 d) S_{10}) + t ((1664 - 276 d + 84 d^2 + 15 d^3) S_3 + (-220 + 78 d - 24 d^2) S_7 \\
&+ (-4 + 3 d) S_{11})) - s t (-6 (-1152 - 440 d + 140 d^2 - 66 d^3 + 5 d^4) S_2 - 2496 S_6 + 1048 d S_6 - 636 d^2 S_6 + 48 d^3 S_6 + 48 S_{10} \\
&+ 68 d S_{10} - 6 d^2 S_{10} + t ((-1424 + 1660 d - 672 d^2 + 111 d^3) S_3 + (132 + 110 d - 48 d^2) S_7 + (8 + 3 d) S_{11}))\Big)\Big) \\
& \Big/ \Big(48 (-2520 + 2754 d - 1175 d^2 + 245 d^3 - 25 d^4 + d^5) s^2 t^2 (s+t) (-m_t^2 + s + t)^2 \Big)
\end{aligned}
$$

$$
\begin{aligned}
B_7 = {}& \Big(32 m_t^6 ((d-2) s + (-7 + d) t) ((4 + 3 d) S_3 - S_7) - 4 m_t^5 \Big((d-2) s (3 (24 + 6 d + d^2) S_1 - 2 (8 + 3 d) S_5 + S_9) \\
&+ t (3 (-168 - 18 d - d^2 + d^3) S_1 - 40 (4 + 3 d) S_4 + 112 S_5 + 26 d S_5 - 6 d^2 S_5 + 40 S_8 - 7 S_9 + d S_9)\Big)
\end{aligned}
$$

$$+ m_t\, t\,(s+t)\, \Big(-4\,t\,(3\,(-168 - 18\,d - d^2 + d^3)\,S_1 + (4 - 74\,d + 36\,d^2)\,S_4 + 112\,S_5 + 26\,d\,S_5 - 6\,d^2\,S_5 + 20\,S_8 - 10\,d\,S_8 - 7\,S_9 + d\,S_9)$$

$$+ s\,(3\,(432 + 132\,d + 24\,d^2 + d^3)\,S_1 + 2\,(-28 + 60\,d - 29\,d^2 + 3\,d^3)\,S_4 - 288\,S_5 - 124\,d\,S_5 - 6\,d^2\,S_5 - 28\,S_8 + 18\,d\,S_8 - 2\,d^2\,S_8 + 18\,S_9$$

$$+ d\,S_9)\Big) + m_t^3\,\Big(4\,(d-2)\,s^2\,(3\,(24 + 6\,d + d^2)\,S_1 - 2\,(8 + 3\,d)\,S_5 + S_9) + 8\,t^2\,(3\,(-168 - 18\,d - d^2 + d^3)\,S_1 + (-78 - 97\,d + 18\,d^2)\,S_4$$

$$+ 112\,S_5 + 26\,d\,S_5 - 6\,d^2\,S_5 + 30\,S_8 - 5\,d\,S_8 - 7\,S_9 + d\,S_9) + s\,t\,(3\,(-1296 - 156\,d - 12\,d^2 + 7\,d^3)\,S_1 + 8\,(-38 - 87\,d + 3\,d^2)\,S_4 + 864\,S_5$$

$$+ 212\,d\,S_5 - 42\,d^2\,S_5 + 160\,S_8 - 54\,S_9 + 7\,d\,S_9)\Big) - t\,(s+t)\,\Big(-(t\,(3\,(-240 + 4\,d + 9\,d^3)\,S_2 + (132 + 110\,d - 48\,d^2)\,S_6 + (-24 + 7\,d)\,S_{10}))$$

$$+ s\,(3\,(32 + 100\,d - 28\,d^2 + 5\,d^3)\,S_2 - 220\,S_6 + 78\,d\,S_6 - 24\,d^2\,S_6 + 4\,S_{10} + 3\,d\,S_{10} + (d-2)\,t\,((172 - 96\,d + 21\,d^2)\,S_3 + (34 - 12\,d)\,S_7 + S_{11}))\Big)$$

$$- 4\,m_t^4\,\Big(8\,(d-2)\,s^2\,((4 + 3\,d)\,S_3 - S_7) + t\,((528 + 24\,d + 42\,d^2 - 3\,d^3)\,S_2 - 24\,S_6 - 80\,d\,S_6 + 6\,d^2\,S_6 + 16\,(-7 + d)\,t\,((4 + 3\,d)\,S_3 - S_7)$$

$$+ 12\,S_{10} - d\,S_{10}) - s\,(3\,(-16 + 12\,d - 4\,d^2 + d^3)\,S_2 - (d-2)\,((-8 + 6\,d)\,S_6 - S_{10}) + t\,((576 + 218\,d - 27\,d^2)\,S_3 + 4\,(-22 + d)\,S_7 + S_{11}))\Big)$$

$$+ m_t^2\,\Big(t^2\,((2832 + 84\,d + 168\,d^2 - 39\,d^3)\,S_2 - 228\,S_6 - 430\,d\,S_6 + 72\,d^2\,S_6 + 32\,(-7 + d)\,t\,((4 + 3\,d)\,S_3 - S_7) + 72\,S_{10} - 11\,d\,S_{10})$$

$$- 2\,s^2\,(6\,(-16 + 12\,d - 4\,d^2 + d^3)\,S_2 - 2\,(d-2)\,((-8 + 6\,d)\,S_6 - S_{10}) + t\,((604 + 72\,d + 71\,d^2 - 3\,d^3)\,S_3 + (-18 - 33\,d + d^2)\,S_7 + 2\,S_{11}))$$

$$+ s\,t\,(3\,(688 + 100\,d + 72\,d^2 - 7\,d^3)\,S_2 - 84\,S_6 - 430\,d\,S_6 + 36\,d^2\,S_6 + 60\,S_{10} - 5\,d\,S_{10} + t\,((-2392 - 444\,d - 126\,d^2 + 21\,d^3)\,S_3$$

$$+ (220 + 74\,d - 12\,d^2)\,S_7 + (-6 + d)\,S_{11}))\Big) \Big)\,\Big/\,\Big(48\,(-2520 + 2754\,d - 1175\,d^2 + 245\,d^3 - 25\,d^4 + d^5)\,s\,t^2\,(s+t)\,(-m_t^2 + s + t)^2 \Big)$$

$$B_8 = \Big(80\,m_t^5\,s\,((4 + 3\,d)\,S_3 - S_7) - (-7 + d)\,s\,(s+t)^2\,(d^2\,(6\,S_1 - 3\,S_4) - 4\,S_4 + 2\,S_5 - 2\,S_8 + d\,(12\,S_1 + 8\,S_4 - 9\,S_5 + S_8) + S_9)$$

$$- 2\,m_t^4\,\Big(-8\,(d-2)\,t\,((4 + 3\,d)\,S_4 - S_8) + s\,(15\,(24 + 6\,d + d^2)\,S_1 - 8\,(-28 - 17\,d + 3\,d^2)\,S_4 - 80\,S_5 - 30\,d\,S_5 - 56\,S_8 + 8\,d\,S_8 + 5\,S_9)\Big)$$

$$+ m_t^2\,(s+t)\,\Big(-16\,(d-2)\,t\,((4 + 3\,d)\,S_4 - S_8) + s\,(6\,(120 + 16\,d + d^3)\,S_1 + (616 + 164\,d - 36\,d^2)\,S_4 - 174\,S_5 + 5\,d\,S_5 - 9\,d^2\,S_5 - 112\,S_8$$

$$+ 16\,d\,S_8 + 3\,S_9 + d\,S_9)\Big) - 2\,m_t^3\,\Big(s^2\,((244 + 66\,d + 6\,d^2)\,S_3 - 40\,S_7) + (d-2)\,t\,(3\,(24 + 6\,d + d^2)\,S_2 - 2\,(8 + 3\,d)\,S_6 + S_{10})$$

$$+ s\,(3\,(-208 - 8\,d - 6\,d^2 + d^3)\,S_2 + 72\,S_6 + 56\,d\,S_6 - 6\,d^2\,S_6 - 12\,S_{10} + d\,S_{10} - t\,((-100 - 116\,d + 9\,d^2)\,S_3 - 10\,(-6 + d)\,S_7 + S_{11}))\Big)$$

$$+ m_t\,(s+t)\,\Big((14 - 9\,d + d^2)\,s^2\,((10 + 3\,d)\,S_3 - S_7) + 2\,(d-2)\,t\,(3\,(24 + 6\,d + d^2)\,S_2 - 2\,(8 + 3\,d)\,S_6 + S_{10}) + 2\,s\,(3\,(-236 - 4\,d + d^2)\,S_2$$

$$+ 114\,S_6 + 29\,d\,S_6 - 3\,d^2\,S_6 - 12\,S_{10} + d\,S_{10} - t\,((60 + 4\,d + 9\,d^2)\,S_3 - 10\,(d-2)\,S_7 + S_{11}))\Big) \Big)$$

$$\Big/\,\Big(24\,(-2520 + 2754\,d - 1175\,d^2 + 245\,d^3 - 25\,d^4 + d^5)\,s^2\,t\,(s+t)\,(-m_t^2 + s + t)^2 \Big)$$

$$B_9 = -\Big(8\,m_t^3\,((d-2)\,s + (-7 + d)\,t)\,((4 + 3\,d)\,S_3 - S_7) + (-7 + d)\,t\,(s+t)\,\Big(3\,(24 + 6\,d + d^2)\,S_1 + (8 - 12\,d)\,S_4 - 16\,S_5 - 6\,d\,S_5 + 4\,S_8 + S_9 \Big)$$

$$- m_t^2\,\Big((d-2)\,s\,(3\,(24 + 6\,d + d^2)\,S_1 - 2\,(8 + 3\,d)\,S_5 + S_9) + t\,(3\,(-168 - 18\,d - d^2 + d^3)\,S_1 - 40\,(4 + 3\,d)\,S_4 + 112\,S_5$$

$$+ 26\,d\,S_5 - 6\,d^2\,S_5 + 40\,S_8 - 7\,S_9 + d\,S_9)\Big) + m_t\,\Big(t\,(3\,(-176 - 8\,d - 14\,d^2 + d^3)\,S_2 + 24\,S_6 + 80\,d\,S_6 - 6\,d^2\,S_6 - 8\,(-7 + d)\,t\,((4 + 3\,d)\,S_3 - S_7)$$

$$- 12\,S_{10} + d\,S_{10}) + s\,(3\,(-16 + 12\,d - 4\,d^2 + d^3)\,S_2 - (d-2)\,((-8 + 6\,d)\,S_6 - S_{10}) + t\,((368 + 86\,d - 9\,d^2)\,S_3 - 2\,(18 + d)\,S_7 + S_{11}))\Big) \Big)$$

$$\Big/\,\Big(96\,(-2520 + 2754\,d - 1175\,d^2 + 245\,d^3 - 25\,d^4 + d^5)\,s\,t^2\,(s+t)\,(-m_t^2 + s + t) \Big)$$

$$B_{10} = \Big(8\,m_t^2\,s\,((d-2)\,s + (-12 + d)\,t)\,((4 + 3\,d)\,S_3 - S_7) + m_t\,\Big(-8\,(d-2)\,t^2\,((4 + 3\,d)\,S_4 - S_8) - (d-2)\,s^2\,(3\,(24 + 6\,d + d^2)\,S_1$$

$$- 2\,(8 + 3\,d)\,S_5 + S_9) - (-12 + d)\,s\,t\,(3\,(24 + 6\,d + d^2)\,S_1 + 8\,(4 + 3\,d)\,S_4 - 16\,S_5 - 6\,d\,S_5 - 8\,S_8 + S_9)\Big)$$

$$+ (d-2)\,t^2\,\Big(3\,(24 + 6\,d + d^2)\,S_2 - 2\,(8 + 3\,d)\,S_6 + S_{10} \Big) - s\,t\,\Big(-6\,(-192 - 8\,d - 10\,d^2 + d^3)\,S_2 + 2\,(-12 + d)\,((4 + 6\,d)\,S_6 - S_{10})$$

$$+ t\,((200 - 142\,d + 27\,d^2)\,S_3 + (48 - 14\,d)\,S_7 + S_{11})\Big) + s^2\,\Big(3\,(-16 + 12\,d - 4\,d^2 + d^3)\,S_2 - (d-2)\,((-8 + 6\,d)\,S_6 - S_{10})$$

$$+ t\,((88 + 42\,d + 3\,d^2)\,S_3 - 2\,(4 + 3\,d)\,S_7 + S_{11})\Big) \Big)\,\Big/\,\Big(96\,(-2520 + 2754\,d - 1175\,d^2 + 245\,d^3 - 25\,d^4 + d^5)\,s^2\,t^2\,(s+t)\,(-m_t^2 + s + t) \Big)$$

$$B_{11} = \Bigg(40\,m_t^2\,s\,((4+3d)\,S_3 - S_7) - 5\,m_t\,\Big(8\,t\,(-((4+3d)\,S_4) + S_8) + s\,(3\,(24+6d+d^2)\,S_1 - 2\,(8+3d)\,S_5 + S_9)\Big)$$

$$- 5\,t\,(3\,(24+6d+d^2)\,S_2 - 2\,(8+3d)\,S_6 + S_{10}) + s\,\Big(15\,(8-2d+d^2)\,S_2 + 5\,((8-6d)\,S_6 + S_{10}) + t\,((144-50d+15d^2)\,S_3$$

$$- 10\,(d-2)\,S_7 + S_{11})\Big)\Bigg) \Big/ \Bigg(480\,(-2520+2754\,d - 1175\,d^2 + 245\,d^3 - 25\,d^4 + d^5)\,s\,t\,(s+t)\,(-m_t^2 + s + t)\Bigg).$$

$$(E.1)$$

Integral families

In the following one can find the integral families belonging to the double box diagrams, shown in tab. 5.2.

- Topology d1
    1) Diagram47

$$\int \frac{d^d k_1}{(2\pi)^d} \frac{d^d k_2}{(2\pi)^d} \frac{1}{(k_1^2 - m_w^2)^{a_1}(k_2^2)^{a_2}((p_1 - k_2)^2)^{a_3}((p_2 - k_1)^2 - m_t^2)^{a_4}}$$
$$\frac{1}{((q_2 - k_1)^2)^{a_5}((p_2 - k_1 + k_2)^2 - m_t^2)^{a_6}((-p_2 + q_1 + k_1 - k_2)^2)^{a_7}((p_1 - k_1)^2)^{a_8}((p_2 + q_2 - k_2)^2)^{a_9}} \tag{F.1}$$

    2) Diagram37

$$\int \frac{d^d k_1}{(2\pi)^d} \frac{d^d k_2}{(2\pi)^d} \frac{1}{(k_2)^{a_1}(k_1^2 - m_w^2)^{a_2}((p_1 - k_1)^2)^{a_3}((p_2 + k_1)^2)^{a_4}((p_1 + p_2 - q_2 + k_2)^2 - m_t^2)^{a_5}}$$
$$\frac{1}{((p_1 - k_1 + k_2)^2 - m_t^2)^{a_6}((-p_1 + q_2 + k_1 - k_2)^2)^{a_7}((k_1 - q_2)^2)^{a_8}((k_1 - k_2 + p_2 - 2p_1 + q_2)^2)^{a_9}} \tag{F.2}$$

- Topology d2
    1) Diagram40

$$\int \frac{d^d k_1}{(2\pi)^d} \frac{d^d k_2}{(2\pi)^d} \frac{1}{(k_2^2)^{a_1}(k_1^2 - m_w^2)^{a_2}((p_1 - k_1)^2)^{a_3}((p_2 - k_2)^2)^{a_4}((q_1 - k_1)^2)^{a_5}}$$
$$\frac{1}{((p_1 - k_1 + k_2)^2)^{a_6}((-p_1 + q_2 + k_1 - k_2)^2)^{a_7}((k_1 + p_2)^2)^{a_8}((k_2 - k_1 + 2p_1 - p_2 - q_2)^2)^{a_9}} \tag{F.3}$$

2) Diagram50

$$\int \frac{d^d k_1}{(2\pi)^d} \frac{d^d k_2}{(2\pi)^d} \frac{1}{(k_1^2 - mw^2)^{a_1}(k_2^2)^{a_2}((p_1 - k_2)^2)^{a_3}((p_1 + p_2 - q_2 + k_1)^2)^{a_4}}$$

$$\frac{1}{((q_2 - k_1)^2)^{a_5}((p_1 + p_2 - q_2 + k_1 - k_2)^2)^{a_6}((p_1 - q_2 + k_1 - k_2)^2)^{a_7}((k_1 - p_2)^2)^{a_8}((k_2 - p_2 - q_2)^2)^{a_9}}$$

$$(\text{F.4})$$

- Topology d3
  1) Diagram35

$$\int \frac{d^d k_1}{(2\pi)^d} \frac{d^d k_2}{(2\pi)^d} \frac{1}{(k_1^2)^{a_1}(k_2^2)^{a_2}((p_1 - k_1)^2)^{a_3}((p_2 + k_1)^2)^{a_4}((q_1 + k_2)^2 - m_t^2)^{a_5}}$$

$$\frac{1}{((p_1 - k_1 + k_2)^2)^{a_6}((p_1 - q_2 - k_1 + k_2)^2 - m_w^2)^{a_7}((k_1 - q_2)^2)^{a_8}((k_2 - p_2)^2)^{a_9}}$$

$$(\text{F.5})$$

2) Diagram46

$$\int \frac{d^d k_1}{(2\pi)^d} \frac{d^d k_2}{(2\pi)^d} \frac{1}{(k_1^2)^{a_1}(k_2^2 - m_w^2)^{a_2}((p_1 - k_2)^2)^{a_3}((p_2 - k_1)^2)^{a_4}((q_2 - k_1)^2)^{a_5}}$$

$$\frac{1}{((p_2 - k_1 + k_2)^2 - m_t^2)^{a_6}((p_1 - q_2 + k_1 - k_2)^2)^{a_7}((-k_1 + p_1)^2)^{a_8}((k_2 - k_1 + q_2 - p_1 - p_2)^2)^{a_9}}$$

$$(\text{F.6})$$

- Topology d4
  1) Diagram36

$$\int \frac{d^d k_1}{(2\pi)^d} \frac{d^d k_2}{(2\pi)^d} \frac{1}{(k_1^2)^{a_1}((k_2^2 - m_w^2))^{a_2}((p_1 - k_1)^2)^{a_3}((p_2 + k_1)^2)^{a_4}((q_1 + k_2)^2)^{a_5}}$$

$$\frac{1}{((p_1 - k_1 + k_2)^2)^{a_6}((p_1 - q_2 - k_1 + k_2)^2)^{a_7}((k_1 - q_2)^2)^{a_8}((k_2 - p_2 - q_2)^2)^{a_9}}$$

$$(\text{F.7})$$

2) Diagram45

$$\int \frac{d^d k_1}{(2\pi)^d} \frac{d^d k_2}{(2\pi)^d} \frac{1}{(k_1^2)^{a_1}(k_2^2)^{a_2}((p_1 - k_2)^2)^{a_3}((p_2 - k_1)^2)^{a_4}((q_2 - k_1)^2)^{a_5}}$$

$$\frac{1}{((p_2 - k_1 + k_2)^2)^{a_6}((p_1 - q_2 + k_1 - k_2)^2 - m_w^2)^{a_7}((p_1 - k_1)^2)^{a_8}((k_2 - k_1 - p_2 + q_2)^2)^{a_9}}$$

$$(\text{F.8})$$

- Topology d5
  1) Diagram38

$$\int \frac{d^d k_1}{(2\pi)^d} \frac{d^d k_2}{(2\pi)^d} \frac{1}{(k_1^2)^{a_1}((k_2^2))^{a_2}((p_1-k_1)^2)^{a_3}((p_2+k_2)^2)^{a_4}((q_1-k_1)^2-m_t^2)^{a_5}}$$

$$\frac{1}{((p_1-k_1+k_2)^2)^{a_6}((p_1-q_2-k_1+k_2)^2-m_w^2)^{a_7}((k_1-q_2)^2)^{a_8}((k_2-p_1-p_2)^2)^{a_9}} \tag{F.9}$$

  2) Diagram49

$$\int \frac{d^d k_1}{(2\pi)^d} \frac{d^d k_2}{(2\pi)^d} \frac{1}{(k_1^2)^{a_1}(k_2^2-m_w^2)^{a_2}((p_1-k_2)^2)^{a_3}((p_1+p_2-q_2+k_1)^2-m_t^2)^{a_4}((q_2-k_1)^2)^{a_5}}$$

$$\frac{1}{((p_1+p_2-q_2+k_1-k_2)^2)^{a_6}((p_1-q_2+k_1-k_2)^2)^{a_7}((k_1-p_1)^2)^{a_8}((k_2-k_1-p_2+2q_2-p_1)^2)^{a_9}} \tag{F.10}$$

- Topology d6
  1) Diagram48

$$\int \frac{d^d k_1}{(2\pi)^d} \frac{d^d k_2}{(2\pi)^d} \frac{1}{(k_1^2)^{a_1}(k_2^2)^{a_2}((p_1-k_2)^2)^{a_3}((q_1+k_1)^2=m_t^2)^{a_4}((q_2-k_1)^2)^{a_5}}$$

$$\frac{1}{((q_1+k_1-k_2)^2-mt^2)^{a_6}((-p_2+q_1+k_1-k_2)^2-m_w^2)^{a_7}((k_1-p_1)^2)^{a_8}((k_2-q_2-p_2)^2)^{a_9}} \tag{F.11}$$

  2) Diagram39

$$\int \frac{d^d k_1}{(2\pi)^d} \frac{d^d k_2}{(2\pi)^d} \frac{1}{(k_1^2)^{a_1}(k_2^2-mw^2)^{a_2}((p_1-k_1)^2)^{a_3}((p_2-k_2)^2-m_t^2)^{a_4}((q_1-k_1)^2-m_t^2)^{a_5}}$$

$$\frac{1}{((p_1-k_1+k_2)^2)^{a_6}((-p_1+q_2+k_1-k_2)^2)^{a_7}((-k_1+q_2)^2)^{a_8}((k_2-k_1-q_2+2p_1-p_2)^2)^{a_9}} \tag{F.12}$$

- Topology d7
  1) Diagram55

$$\int \frac{d^d k_1}{(2\pi)^d} \frac{d^d k_2}{(2\pi)^d} \frac{1}{(k_1^2)^{a_1}(k_2^2-m_w^2)^{a_2}((p_1-k_1)^2)^{a_3}((p_2+k_2)^2-m_t^2)^{a_4}((q_1-k_1)^2-m_t^2)^{a_5}}$$

$$\frac{1}{((q_2+k_2)^2)^{a_6}((-p_1+q_2+k_1+k_2)^2)^{a_7}((k_1-q_2)^2)^{a_8}((k_2-p_1-q_2)^2)^{a_9}} \tag{F.13}$$

2) Diagram53

$$\int \frac{d^d k_1}{(2\pi)^d} \frac{d^d k_2}{(2\pi)^d} \frac{1}{(k_2^2)^{a_1}(k_1^2 - m_w^2)^{a_2}((p_1 - k_1)^2)^{a_3}(((p_2 + k_1)^2 - m_t^2)^{a_4}((q_1 - k_2)^2 - m_t^2)^{a_5}}$$
$$\frac{1}{((q_2 + k_2)^2)^{a_6}((q_2 - p_1 + k_1 + k_2)^2)^{a_7}((k_2 + p_1)^2)^{a_8}((k_1 + q_2 + p_1)^2)^{a_9}} \tag{F.14}$$

- Topology d8
  1) Diagram56

$$\int \frac{d^d k_1}{(2\pi)^d} \frac{d^d k_2}{(2\pi)^d} \frac{1}{(k_2^2)^{a_1}(k_1^2 - m_w^2)^{a_2}((p_1 - k_1)^2)^{a_3}((p_2 + k_2)^2)^{a_4}((q_1 - k_1)^2)^{a_5}}$$
$$\frac{1}{((q_2 + k_2)^2)^{a_6}((-p_1 + q_2 + k_1 + k_2)^2)^{a_7}((k_1 - p_2)^2)^{a_8}((k_2 - p_1 - q_2)^2)^{a_9}} \tag{F.15}$$

2) Diagram52

$$\int \frac{d^d k_1}{(2\pi)^d} \frac{d^d k_2}{(2\pi)^d} \frac{1}{(k_1^2)^{a_1}(k_2^2 - m_w^2)^{a_2}((p_1 - k_1)^2)^{a_3}((p_2 + k_1)^2)^{a_4}((q_1 - k_2)^2)^{a_5}}$$
$$\frac{1}{((q_2 + k_2)^2)^{a_6}((p_1 - q_2 - k_1 - k_2)^2)^{a_7}((k_2 - p_2)^2)^{a_8}((k_1 + p_1 + q_2)^2)^{a_9}} \tag{F.16}$$

- Topology d9
  1) Diagram54

$$\int \frac{d^d k_1}{(2\pi)^d} \frac{d^d k_2}{(2\pi)^d} \frac{1}{(k_1^2)^{a_1}(k_2^2)^{a_2}((p_1 - k_1)^2)^{a_3}((p_2 + k_2)^2)^{a_4}((q_1 - k_1)^2 - m_t^2)^{a_5}}$$
$$\frac{1}{((q_2 + k_2)^2)^{a_6}((-p_1 + q_2 + k_1 + k_2)^2 - m_w^2)^{a_7}((k_1 - p_2)^2)^{a_8}((k_2 - p_1 - q_2)^2)^{a_9}} \tag{F.17}$$

2) Diagram51

$$\int \frac{d^d k_1}{(2\pi)^d} \frac{d^d k_2}{(2\pi)^d} \frac{1}{(k_1^2)^{a_1}(k_2^2)^{a_2}((p_1 - k_1)^2)^{a_3}((p_2 + k_1)^2)^{a_4}((q_1 - k_2)^2 - m_t^2)^{a_5}}$$
$$\frac{1}{(((q_2 + k_2)^2)^{a_6}((p_1 - q_2 - k_1 - k_2)^2 - m_w^2)^{a_7}((k_2 - p_2)^2)^{a_8}((k_1 + p_1 + q_2)^2)^{a_9}} \tag{F.18}$$

[1] **A**TLAS Collaboration Collaboration, W. Armstrong *et. al.*, *ATLAS: Technical proposal for a general-purpose p p experiment at the Large Hadron Collider at CERN,* .

[2] **C**MS Collaboration Collaboration, *CMS, the Compact Muon Solenoid: Technical proposal,* .

[3] W. Bernreuther, R. Bonciani, T. Gehrmann, R. Heinesch, T. Leineweber, *et. al.*, *Two-loop QCD corrections to the heavy quark form-factors: Axial vector contributions,* Nucl.Phys. **B**712 (2005) 229–286, [hep-ph/0412259].

[4] K. G.Chetyrkin and F. V. Tkachov *Nucl. Phys.* **B**192 ((1981)) 159.

[5] R. Lee, *Space-time dimensionality D as complex variable: Calculating loop integrals using dimensional recurrence relation and analytical properties with respect to D, Nucl.Phys.* **B**830 (2010) 474–492, [arXiv:0911.0252].

[6] **C**MS Collaboration Collaboration, S. Chatrchyan *et. al.*, *Observation of a new boson at a mass of 125 GeV with the CMS experiment at the LHC, Phys.Lett.* **B**716 (2012) 30–61, [arXiv:1207.7235].

[7] **A**TLAS Collaboration Collaboration, G. Aad *et. al.*, *Observation of a new particle in the search for the Standard Model Higgs boson with the ATLAS detector at the LHC, Phys.Lett.* **B**716 (2012) 1–29, [arXiv:1207.7214].

[8] D. Gross and F. Wilczek, *Ultraviolet Behavior of Nonabelian Gauge Theories, Phys.Rev.Lett.* **30** (1973) 1343–1346.

[9] **C**TEQ Collaboration Collaboration, R. Brock *et. al.*, *Handbook of perturbative QCD: Version 1.0, Rev.Mod.Phys.* **67** (1995) 157–248.

[10] R. K. Ellis, W. J. Stirling, and B. Webber, *QCD and collider physics, Camb.Monogr.Part.Phys.Nucl.Phys.Cosmol.* **8** (1996) 1–435.

[11] NLO Multileg Working Group Collaboration, Z. Bern *et. al.*, *The NLO multileg working group: Summary report*, arXiv:0803.0494.

[12] E. N. Glover, *Progress in NNLO calculations for scattering processes*, *Nucl.Phys.Proc.Suppl.* **116** (2003) 3–7, [hep-ph/0211412].

[13] R. V. Harlander and W. B. Kilgore, *Next-to-next-to-leading order Higgs production at hadron colliders*, *Phys.Rev.Lett.* **88** (2002) 201801, [hep-ph/0201206].

[14] B. Harris, E. Laenen, L. Phaf, Z. Sullivan, and S. Weinzierl, *The Fully differential single top quark cross-section in next to leading order QCD*, *Phys.Rev.* **D66** (2002) 054024, [hep-ph/0207055].

[15] G. Mahlon and S. J. Parke, *Improved spin basis for angular correlation studies in single top quark production at the Tevatron*, *Phys.Rev.* **D55** (1997) 7249–7254, [hep-ph/9611367].

[16] G. Mahlon and S. J. Parke, *Single top quark production at the LHC: Understanding spin*, *Phys.Lett.* **B476** (2000) 323–330, [hep-ph/9912458].

[17] Higgs Working Group Collaboration Collaboration, M. S. Carena *et. al.*, *Report of the Tevatron Higgs working group*, hep-ph/0010338.

[18] S. Jabeen, *Top and Higgs Physics at the Hadron Colliders*, *Int.J.Mod.Phys.* **A28** (2013) 1330038.

[19] SUGRA Working Group Collaboration Collaboration, V. Barger *et. al.*, *Report of the SUGRA Working Group for run II of the Tevatron*, hep-ph/0003154.

[20] R parity Working Group Collaboration Collaboration, B. Allanach *et. al.*, *Searching for R parity violation at Run II of the Tevatron*, hep-ph/9906224.

[21] A. Onofre, *Searches for new physics in top decays at the LHC*, arXiv:1212.6898.

[22] ATLAS Collaboration Collaboration, G. Aad *et. al.*, *Search for FCNC single top-quark production at $\sqrt{s} = 7$ TeV with the ATLAS detector*, *Phys.Lett.* **B712** (2012) 351–369, [arXiv:1203.0529].

[23] Collaboration for the ATLAS, CDF Collaboration, CMS Collaboration, D0 Collaboration Collaboration, E. Yazgan, *Flavor changing neutral currents in top quark production and decay*, arXiv:1312.5435.

[24] S. Laporta, *High precision calculation of multiloop Feynman integrals by difference equations*, *Int.J.Mod.Phys.* **A15** (2000) 5087–5159, [hep-ph/0102033].

[25] A. von Manteuffel and C. Studerus, *Reduze 2 - Distributed Feynman Integral Reduction*, arXiv:1201.4330.

[26] G. Passarino and M. Veltman, *One Loop Corrections for e+ e- Annihilation Into mu+ mu- in the Weinberg Model*, *Nucl.Phys.* **B160** (1979) 151.

[27] I. Binder, (ed.) and D. Kreimer, (ed.), *Universality and renormalization: From stochastic evolution to renormalization of quantum fields. Proceedings, Workshop on 'Percolation, SLE and related topics', Toronto, Canada, September 20-24, 2005, and Workshop on 'Renormaliza, .* Prepared for Workshop on Renormalization and Universality in Mathematical Physics, Toronto, Canada, 18-22 Oct 2005.

[28] A. Denner and S. Dittmaier, *Reduction schemes for one-loop tensor integrals, Nucl.Phys.* **B**734 (2006) 62–115, [hep-ph/0509141].

[29] R. K. Ellis, Z. Kunszt, K. Melnikov, and G. Zanderighi, *One-loop calculations in quantum field theory: from Feynman diagrams to unitarity cuts, Phys.Rept.* **518** (2012) 141–250, [arXiv:1105.4319].

[30] O. Tarasov, *Generalized recurrence relations for two loop propagator integrals with arbitrary masses, Nucl.Phys.* **B**502 (1997) 455–482, [hep-ph/9703319].

[31] O. Tarasov, *Connection between Feynman integrals having different values of the space-time dimension, Phys.Rev.* **D**54 (1996) 6479–6490, [hep-th/9606018].

[32] C. Anastasiou, E. N. Glover, and C. Oleari, *The two loop scalar and tensor pentabox graph with lightlike legs, Nucl.Phys.* **B**575 (2000) 416–436, [hep-ph/9912251].

[33] E. N. Glover, *Two loop QCD helicity amplitudes for massless quark quark scattering, JHEP* **0404** (2004) 021, [hep-ph/0401119].

[34] C. Anastasiou, E. N. Glover, C. Oleari, and M. Tejeda-Yeomans, *Two-loop QCD corrections to the scattering of massless distinct quarks, Nucl.Phys.* **B**601 (2001) 318–340, [hep-ph/0010212].

[35] C. Anastasiou, E. N. Glover, C. Oleari, and M. Tejeda-Yeomans, *Two loop QCD corrections to massless identical quark scattering, Nucl.Phys.* **B**601 (2001) 341–360, [hep-ph/0011094].

[36] T. Gehrmann and E. Remiddi, *Differential equations for two loop four point functions, Nucl.Phys.* **B**580 (2000) 485–518, [hep-ph/9912329].

[37] A. Smirnov and V. Smirnov, *S-bases as a tool to solve reduction problems for Feynman integrals, Nucl.Phys.Proc.Suppl.* **160** (2006) 80–84, [hep-ph/0606247].

[38] P. Baikov, *The Criterion of irreducibility of multiloop Feynman integrals, Phys.Lett.* **B**474 (2000) 385–388, [hep-ph/9912421].

[39] P. Baikov, *Explicit solutions of the multiloop integral recurrence relations and its application, Nucl.Instrum.Meth.* **A**389 (1997) 347–349, [hep-ph/9611449].

[40] P. Baikov, *Explicit solutions of n loop vacuum integral recurrence relations,* hep-ph/9604254.

[41] P. Baikov, *A Practical criterion of irreducibility of multi-loop Feynman integrals, Phys.Lett.* **B**634 (2006) 325–329, [hep-ph/0507053].

[42] B. Buchberger and F. W. (eds.), *Gröbner* bases and applications, .

[43] V. A. Smirnov, *Evaluating Feynman integrals*. Springer tracts in modern physics. Springer, Berlin, 2004.

[44] O. Tarasov, *Reduction of Feynman graph amplitudes to a minimal set of basic integrals*, *Acta Phys.Polon.* **B29** (1998) 2655, [hep-ph/9812250].

[45] S. Laporta and E. Remiddi, *The Analytical value of the electron (g-2) at order alpha**3 in QED*, *Phys.Lett.* **B379** (1996) 283–291, [hep-ph/9602417].

[46] C. Anastasiou and A. Lazopoulos, *Automatic integral reduction for higher order perturbative calculations*, *JHEP* **0407** (2004) 046, [hep-ph/0404258].

[47] A. Smirnov, *Algorithm FIRE – Feynman Integral REduction*, *JHEP* **0810** (2008) 107, [arXiv:0807.3243].

[48] C. Studerus, *Reduze-Feynman Integral Reduction in C++*, *Comput.Phys.Commun.* **181** (2010) 1293–1300, [arXiv:0912.2546].

[49] C. Bogner and S. Weinzierl, *Feynman graph polynomials*, *Int.J.Mod.Phys.* **A25** (2010) 2585–2618, [arXiv:1002.3458].

[50] T. Binoth and G. Heinrich, *An Automatized algorithm to compute infrared divergent multiloop integrals*, *Nucl.Phys.* **B585** (2000) 741–759, [hep-ph/0004013].

[51] S. Weinzierl, *The Art of computing loop integrals*, hep-ph/0604068.

[52] C. Itzykson and J. Zuber, *Quantum field theory: Claude Itzykson and Jean-Bernard Zuber*. Dover Books on Physics Series. McGraw-Hill, 2005.

[53] E. Speer, *Renormalization and ward identities using complex space-time dimension*, *J.Math.Phys.* **15** (1974) 1–6.

[54] A. Smirnov and V. Smirnov, *On the reduction of Feynman integrals to master integrals*, *PoS* **ACAT2007** (2007) 085, [arXiv:0707.3993].

[55] P. Baikov and V. A. Smirnov, *Equivalence of recurrence relations for Feynman integrals with the same total number of external and loop momenta*, *Phys.Lett.* **B477** (2000) 367–372, [hep-ph/0001192].

[56] R. N. Lee and V. A. Smirnov, *The Dimensional Recurrence and Analyticity Method for Multicomponent Master Integrals: Using Unitarity Cuts to Construct Homogeneous Solutions*, arXiv:1209.0339.

[57] P. Baikov, *Explicit solutions of the three loop vacuum integral recurrence relations*, *Phys.Lett.* **B385** (1996) 404–410, [hep-ph/9603267].

[58] R. Lee, A. Smirnov, and V. Smirnov, *Analytic Results for Massless Three-Loop Form Factors*, *JHEP* **1004** (2010) 020, [arXiv:1001.2887].

[59] R. Lee, *Group structure of the integration-by-part identities and its application to the reduction of multiloop integrals*, JHEP **0807** (2008) 031, [arXiv:0804.3008].

[60] A. Grozin, *Integration by parts: An Introduction*, Int.J.Mod.Phys. **A26** (2011) 2807–2854, [arXiv:1104.3993].

[61] W. Research, *Mathematica Edition: Version 8.0, Champaign, Illinois* (2010) [http://www.wolfram.com/mathematica/].

[62] G. A. Gratzer, *Lattice theory : first concepts and distributive lattices.* A Series of books in mathematics. W. H. Freeman, San Francisco, 1971.

[63] W. Bernreuther, *Top quark physics at the LHC*, J.Phys. **G35** (2008) 083001, [arXiv:0805.1333].

[64] F.-P. Schilling, *Top Quark Physics at the LHC: A Review of the First Two Years*, Int.J.Mod.Phys. **A27** (2012) 1230016, [arXiv:1206.4484].

[65] T. M. Liss, *Top Quark Properties*, arXiv:1212.0489.

[66] **D**0 Collaboration Collaboration, S. Abachi *et. al.*, *Search for high mass top quark production in $p\bar{p}$ collisions at $\sqrt{s} = 1.8$ TeV*, Phys.Rev.Lett. **74** (1995) 2422–2426, [hep-ex/9411001].

[67] **D**0 Collaboration Collaboration, V. Abazov *et. al.*, *Evidence for production of single top quarks and first direct measurement of | Vtb |*, Phys.Rev.Lett. **98** (2007) 181802, [hep-ex/0612052].

[68] **D**0 Collaboration Collaboration, V. Abazov *et. al.*, *Evidence for production of single top quarks*, Phys.Rev. **D78** (2008) 012005, [arXiv:0803.0739].

[69] **CDF** Collaboration Collaboration, T. Aaltonen *et. al.*, *First Observation of Electroweak Single Top Quark Production*, Phys.Rev.Lett. **103** (2009) 092002, [arXiv:0903.0885].

[70] **D**0 Collaboration Collaboration, V. Abazov *et. al.*, *Observation of Single Top Quark Production*, Phys.Rev.Lett. **103** (2009) 092001, [arXiv:0903.0850].

[71] **ATLAS** Collaboration Collaboration, G. Aad *et. al.*, *Measurement of the t-channel single top-quark production cross section in pp collisions at $\sqrt{s} = 7$ TeV with the ATLAS detector*, Phys.Lett. **B717** (2012) 330–350, [arXiv:1205.3130].

[72] **CMS** Collaboration Collaboration, S. Chatrchyan *et. al.*, *Measurement of the single-top-quark t-channel cross section in pp collisions at $\sqrt{s} = 7$ TeV*, JHEP **1212** (2012) 035, [arXiv:1209.4533].

[73] N. Kidonakis, *Next-to-next-to-leading-order collinear and soft gluon corrections for t-channel single top quark production*, Phys.Rev. **D83** (2011) 091503, [arXiv:1103.2792].

[74] G. Chiarelli, *Single Top Physics at Hadron Colliders*, EPJ Web Conf. **49** (2013) 04004, [arXiv:1302.1773].

[75] M. Jezabek and J. H. Kuhn, *V-A tests through leptons from polarized top quarks*, *Phys.Lett.* **B329** (1994) 317–324, [hep-ph/9403366].

[76] J. van der Heide, E. Laenen, L. Phaf, and S. Weinzierl, *Helicity amplitudes for single top production*, *Phys.Rev.* **D62** (2000) 074025, [hep-ph/0003318].

[77] D. Espriu and J. Manzano, *A Study of top polarization in single top production at the CERN LHC*, *Phys.Rev.* **D66** (2002) 114009, [hep-ph/0209030].

[78] N. Kidonakis, *NNLL resummation for s-channel single top quark production*, *Phys.Rev.* **D81** (2010) 054028, [arXiv:1001.5034].

[79] N. Kidonakis, *Two-loop soft anomalous dimensions for single top quark associated production with a W- or H-*, *Phys.Rev.* **D82** (2010) 054018, [arXiv:1005.4451].

[80] H. X. Zhu, C. S. Li, J. Wang, and J. J. Zhang, *Factorization and resummation of s-channel single top quark production*, *JHEP* **1102** (2011) 099, [arXiv:1006.0681].

[81] N. Kidonakis, *Top Quark Production*, arXiv:1311.0283.

[82] Q.-H. Cao, R. Schwienhorst, J. A. Benitez, R. Brock, and C.-P. Yuan, *Next-to-leading order corrections to single top quark production and decay at the Tevatron: 2. t^- channel process*, *Phys.Rev.* **D72** (2005) 094027, [hep-ph/0504230].

[83] Q.-H. Cao, R. Schwienhorst, and C.-P. Yuan, *Next-to-leading order corrections to single top quark production and decay at Tevatron. 1. s^- channel process*, *Phys.Rev.* **D71** (2005) 054023, [hep-ph/0409040].

[84] T. Stelzer, Z. Sullivan, and S. Willenbrock, *Single top quark production via W - gluon fusion at next-to-leading order*, *Phys.Rev.* **D56** (1997) 5919–5927, [hep-ph/9705398].

[85] T. Stelzer, Z. Sullivan, and S. Willenbrock, *Single top quark production at hadron colliders*, *Phys.Rev.* **D58** (1998) 094021, [hep-ph/9807340].

[86] M. C. Smith and S. Willenbrock, *QCD and Yukawa corrections to single top quark production via $q\bar{q} \to t\bar{b}$*, *Phys.Rev.* **D54** (1996) 6696–6702, [hep-ph/9604223].

[87] P. Nogueira, *Automatic Feynman graph generation*, *J.Comput.Phys.* **105** (1993) 279–289.

[88] S. L. Adler, *Axial-vector vertex in spinor electrodynamics*, *Phys. Rev.* **177** (Jan, 1969) 2426–2438.

[89] J. Korner, D. Kreimer, and K. Schilcher, *A Practicable gamma(5) scheme in dimensional regularization*, *Z.Phys.* **C54** (1992) 503–512.

[90] D. Kreimer, *The $\gamma(5)$ Problem and Anomalies: A Clifford Algebra Approach*, *Phys.Lett.* **B237** (1990) 59.

[91] D. Kreimer, *The Role of gamma(5) in dimensional regularization*, hep-ph/9401354.

[92] S. Larin, *The Renormalization of the axial anomaly in dimensional regularization*, *Phys.Lett.* **B303** (1993) 113–118, [hep-ph/9302240].

[93] A. Denner, *Techniques for calculation of electroweak radiative corrections at the one loop level and results for W physics at LEP-200, Fortsch.Phys.* **41** (1993) 307–420, [arXiv:0709.1075].

[94] W. Bernreuther, R. Bonciani, T. Gehrmann, R. Heinesch, T. Leineweber, *et. al.*, *Two-loop QCD corrections to the heavy quark form-factors: The Vector contributions, Nucl.Phys.* **B706** (2005) 245–324, [hep-ph/0406046].

[95] J. Gluza, A. Mitov, S. Moch, and T. Riemann, *The QCD form factor of heavy quarks at NNLO, JHEP* **0907** (2009) 001, [arXiv:0905.1137].

[96] LEP Collaboration, ALEPH Collaboration, DELPHI Collaboration, L3 Collaboration, OPAL Collaboration, LEP Electroweak Working Group, SLD Electroweak Group, SLD Heavy Flavor Group Collaboration, t. S. Electroweak, *A Combination of preliminary electroweak measurements and constraints on the standard model*, hep-ex/0312023.

[97] A. Freitas and K. Monig, *Corrections to quark asymmetries at LEP, Eur.Phys.J.* **C40** (2005) 493, [hep-ph/0411304].

[98] W. Bernreuther, R. Bonciani, T. Gehrmann, R. Heinesch, T. Leineweber, *et. al.*, *Two-loop QCD corrections to the heavy quark form-factors: Anomaly contributions, Nucl.Phys.* **B723** (2005) 91–116, [hep-ph/0504190].

[99] G. 't Hooft and M. Veltman, *Regularization and Renormalization of Gauge Fields, Nucl.Phys.* **B44** (1972) 189–213.

[100] T. Gehrmann, T. Huber, and D. Maitre, *Two-loop quark and gluon form-factors in dimensional regularisation, Phys.Lett.* **B622** (2005) 295–302, [hep-ph/0507061].

[101] R. Bonciani and A. Ferroglia, *Two-Loop QCD Corrections to the Heavy-to-Light Quark Decay, JHEP* **0811** (2008) 065, [arXiv:0809.4687].

[102] M. Beneke, T. Huber, and X.-Q. Li, *Two-loop QCD correction to differential semi-leptonic b —gt; u decays in the shape-function region, Nucl.Phys.* **B811** (2009) 77–97, [arXiv:0810.1230].

[103] H. Asatrian, C. Greub, and B. Pecjak, *NNLO corrections to anti-B —gt; X(u) l anti-nu in the shape-function region, Phys.Rev.* **D78** (2008) 114028, [arXiv:0810.0987].

[104] G. Bell, *NNLO corrections to inclusive semileptonic B decays in the shape-function region, Nucl.Phys.* **B812** (2009) 264–289, [arXiv:0810.5695].

[105] G. Bell, *Higher order QCD corrections in exclusive charmless B decays*, arXiv:0705.3133.

[106] R. Bonciani, P. Mastrolia, and E. Remiddi, *Master integrals for the two loop QCD virtual corrections to the forward backward asymmetry, Nucl.Phys.* **B690** (2004) 138–176, [hep-ph/0311145].

[107] R. Bonciani, A. Ferroglia, T. Gehrmann, D. Maitre, and C. Studerus, *Two-Loop Fermionic Corrections to Heavy-Quark Pair Production: The Quark-Antiquark Channel*, JHEP **0807** (2008) 129, [arXiv:0806.2301].

[108] J. Fleischer, M. Y. Kalmykov, and A. Kotikov, *Two loop selfenergy master integrals on-shell*, Phys.Lett. **B462** (1999) 169–177, [hep-ph/9905249].

[109] M. Argeri, P. Mastrolia, and E. Remiddi, *The Analytic value of the sunrise selfmass with two equal masses and the external invariant equal to the third squared mass*, Nucl.Phys. **B631** (2002) 388–400, [hep-ph/0202123].

[110] U. Aglietti and R. Bonciani, *Master integrals with one massive propagator for the two loop electroweak form-factor*, Nucl.Phys. **B668** (2003) 3–76, [hep-ph/0304028].

[111] J. Fleischer, A. Kotikov, and O. Veretin, *Analytic two loop results for selfenergy type and vertex type diagrams with one nonzero mass*, Nucl.Phys. **B547** (1999) 343–374, [hep-ph/9808242].

[112] S. Buehler and C. Duhr, *CHAPLIN - Complex Harmonic Polylogarithms in Fortran*, arXiv:1106.5739.

[113] R. H. Lewis, *Fermat — shareware program* , http://home.bway.net/lewis/.

[114] M. W. Robert H. Lewis, *Comparison of Polynomial-Oriented Computer Algebra Systems, ISSAC Conference, Vancouver, British Columbia* (28-31 July 1999) [http://home.bway.net/lewis/calatex.html].

[115] C. W. Bauer, A. Frink, and R. Kreckel, *Introduction to the GiNaC framework for symbolic computation within the C++ programming language*, cs/0004015.

[116] Particle Data Group Collaboration, J. Beringer *et. al.*, *Review of Particle Physics (RPP)*, Phys.Rev. **D86** (2012) 010001.

[117] CDF Collaboration, D0 Collaboration Collaboration, T. Aaltonen *et. al.*, *Combination of the top-quark mass measurements from the Tevatron collider*, Phys.Rev. **D86** (2012) 092003, [arXiv:1207.1069].

[118] A. Kotikov, *Differential equations method: New technique for massive Feynman diagrams calculation*, Phys.Lett. **B254** (1991) 158–164.

[119] E. Remiddi, *Differential equations for Feynman graph amplitudes*, Nuovo Cim. **A**110 (1997) 1435–1452, [hep-th/9711188].

[120] M. Czakon, *Tops from Light Quarks: Full Mass Dependence at Two-Loops in QCD*, Phys.Lett. **B664** (2008) 307–314, [arXiv:0803.1400].

[121] P. Baernreuther, M. Czakon, and P. Fiedler, *Virtual amplitudes and threshold behaviour of hadronic top-quark pair-production cross sections*, JHEP **1402** (2014) 078, [arXiv:1312.6279].

[122] L. Fletcher, H. Hudson, G. Cauzzi, K. Getman, M. Giampapa, *et. al.*, *Splinter Session 'Solar and Stellar Flares'*, arXiv:1206.3997.

[123] Z. Sullivan, *Understanding single-top-quark production and jets at hadron colliders*, *Phys.Rev.* **D70** (2004) 114012, [hep-ph/0408049].

[124] P. W. Higgs, *Broken Symmetries and the Masses of Gauge Bosons*, *Phys.Rev.Lett.* **13** (1964) 508–509.

[125] G. 't Hooft, *Renormalizable Lagrangians for Massive Yang-Mills Fields*, *Nucl.Phys.* **B35** (1971) 167–188.

[126] T. Gehrmann, L. Tancredi, and E. Weihs, *Two-loop master integrals for $q\bar{q} \to VV$: the planar topologies*, *JHEP* **1308** (2013) 070, [arXiv:1306.6344].

[127] J. M. Henn, K. Melnikov, and V. A. Smirnov, *Two-loop planar master integrals for the production of off-shell vector bosons in hadron collisions*, arXiv:1402.7078.

[128] A.-S. Schade, *Produktion einzelner Top-Quarks in nächst-zu-nächstführender Ordnung der QCD: Der Beitrag der Einschleifen amplituden quadriert*, arXiv:1310.5725.

[129] J. C. Collins, *Renormalization: an introduction to renormalization, the renormalization group, and the operator-product expansion*. Cambridge monographs on mathematical physics. Cambridge Univ. Press, Cambridge, 1984.

Danksagung

An erster Stelle möchte ich mich bei meinem Doktorvater Peter Uwer dafür bedanken, dieses Projekt ermöglicht und die Betreuung meiner Arbeit übernommen zu haben. Ihm danke ich auch für seine Unterstüzung und das beständige Interesse am Fortgang der Arbeit.

Weiterhin danke ich Sven-Olaf Moch für die hilfsreichen Gespräche nicht nur über das Projekt sondern auch über allgemeinere Themen aus dem Standardmodell.

Mein größter Dank gilt Bas Tausk. In allen unseren Disskusionen konnte er oftmals mir weiterhelfen. Seine geduldige Art und sein unerschöpfliches Wissen über die Quantenfeldtheorie waren immer wegweisend. Es ist mir deshalb ein besonderes Anliegen mich an dieser Stelle nochmals bei Bas zu bedanken.

Bedanken möchte ich mich auch bei der ganzen Arbeitsgruppe "Phänomenologie der Elementarteilchen" für Hilfen, nette Gespräche aber auch fachliche Disskusionen. Ein besonderer Dank geht an Benedikt Biedermann, Peter Galler, Sophia Schade.

Ganz herzlich möchte ich mich bei Desislava Chetalova, Dominik Schneider, Peter Galler, Ralf Sattler und Bas Tausk bedanken, Teile dieser Arbeit Korrektur gelesen zu haben und somit haben sie einen wichtigen Beitrag zur Verbesserung dieser Arbeit.

Ein ganz lieber Dank geht an Desislava Chetalova, Benedikt Biedermann, Stefan Weinzierl und Martin Reuter für die aufheiternden Gespräche in dieser Zeit.

Ich danke auch allen Verantwortlichen des Graduiertenkollegs 1504 "Masse Spektrum Symmetrie" der deutschen Forschungsgemeinschaft für die Finanzierung meiner Doktorarbeit, insbesondere danke ich Sylvia Richter und Martin zur Nedden und auch in diesem Zusammenhang gilt mein Dank dem Sonderforschungsbereich "SFB/TR9 Computergestützte Theoretische Teilchenphysik".

Für interessante Diskussionen, die zu neuen Ideen in der Rechnung geführt haben, danke ich ferner Thomas Gehrmann, Dominik Stöckinger und Andrey Grozin so wie Cedric Studerus.